Practical Information For Beginners in Bee-Keeping

by Wilmon Newell

with an introduction by Jackson Chambers

This work contains material that was originally published in 1911.

This publication is within the Public Domain.

This edition is reprinted for educational purposes and in accordance with all applicable Federal Laws.

Introduction Copyright 2018 by Jackson Chambers

COVER CREDITS

Front Cover
Apis mellifera Western honey bee by Andreas Trepte (Own work)
[CC BY-SA 2.5 - https://creativecommons.org/licenses/by-sa/2.5],
via Wikimedia Commons

Back Cover
Honey-benefits by Lama Raheem (Own work)
[CC BY-SA 4.0 - https://creativecommons.org/licenses/by-sa/4.0],
via Wikimedia Commons

Research / Resources
The Top 6 Raw Honey Benefits
https://www.healthline.com/health/food-nutrition/top-raw-honey-benefits#1
via HealthLine.com

Wikimedia Commons
www.Commons.Wikimedia.org

Many thanks to all the incredible photographers, artists,
researchers, and archivists who share their great work.

PLEASE NOTE :
As with all reprinted books of this age that are intended to perfectly reproduce the original edition, considerable pains and effort had to be undertaken to correct fading and sometimes outright damage to existing proofs of this title. At times, this task can be quite monumental, requiring an almost total rebuilding of some pages from digital proofs of multiple copies. Despite this, imperfections still sometimes exist in the final proof and may detract slightly from the visual appearance of the text.

DISCLAIMER :
Due to the age of this book, some methods or practices may have been deemed unsafe or unacceptable in the interim years. In utilizing the information herein, you do so at your own risk. We republish antiquarian books without judgment or revisionism, solely for their historical and cultural importance, and for educational purposes.

Self Reliance Books

Get more historic titles on animal and stock breeding, gardening and old fashioned skills by visiting us at:

http://selfreliancebooks.blogspot.com/

introduction

Here at **Self-Reliance Books** we are dedicated to bringing you the best in *dusty-old-book-knowledge* to help you in your quest for self-sufficiency and food independence. We're so pleased to bring you this wonderful old title on Apiculture.

Not only is raw honey sweet and divine-tasting, it also has many health benefits including antiviral and antibacterial actions, wound-healing properties, and is packed with phytonutrients and antioxidants.

This special edition of **Practical Information for Beginners in Bee-Keeping** was written by Wilmon Newell, and first published in 1911 making it over a century old.

The book has sections on *Desirability of Keeping Bees, Life History of the Honey Bee, Hives, Location of Hives and Apiary, Wintering, Diseases of the Bees,* and more.

A super-short, fast read, and a great book to start with for all those starting out in Bee-Keeping, or considering taking the plunge.

~ *Roger Chambers*
State of Jefferson, March 2018

English: Bee macro
Date 18 July 2015, 10:35:52
Source https://www.flickr.com/photos/55293400@N07/20118812536/
Author Ömer Ünlü
Wikipedia.com - Creative Commons Attribution 2.0 Generic

Description: Honey bee
Date: 4 November 2011, 12:55
Source: Honey bee
Author AJC1 from UK
Wikipedia.com - Creative Commons Attribution 2.0 Generic

CONTENTS

Introduction	5
Desirability of Keeping Bees	5
How to Secure Bees	7
Life History of the Honey Bee	8
Hives	11
Other Equipment	14
How to Open a Hive	15
Location of Hives and Apairy	17
Equipment for Extensive Work	18
Transferring	21
Robbing	23
Wintering	24
Feeding	24
Honey Flows	24
Cultivated Honey Plants	26
Supers	26
Marketing Honey	28
Fraudulent Packing	28
Beeswax	29
Swarming	29
Prevention and Control of Swarming	30
Uniting Weak Colonies	33
Queen-Rearing	34
Laying Workers	35
Enemies of Bees	35
The Wax-worm	36
Ants and Other Enemies	37
Birds	38
Diseases of Bees	39
American Foul Brood	39
State Control and Eradication of Foul Brood	39
Treatment for American Foul Brood	40
European Foul Brood	43
Pickled Brood	43
Chilled, Starved, Over-heated Brood, etc.	44
Paralysis	45
Educational	46

PRACTICAL INFORMATION FOR BEGINNERS IN BEEKEEPING

BY WILMON NEWELL,

State Entomologist and Entomologist of the Texas Experiment Station.

The present paper is a revision of a bulletin published by the State Entomologist's office at College Station in 1902, the edition of which has been exhausted for some time. The constant demand from farmers, fruit growers and others for information which will enable them to keep bees on a small scale makes a new issue imperative. This paper, like the original edition, is not intended for professional beekeepers or specialists, but for those who possess little or no knowledge concerning bees, and wish to learn how to care for a few colonies. For that reason technical discussions are strictly avoided.

By law, the Professor of Entomology at the Agricultural and Mechanical College of Texas is made State Entomologist, and as such is charged with enforcement of the law for control of diseases of bees and with maintaining an experimental apiary in which experiments are conducted for the benefit of Texas beekeepers. Several colonies of pure-bred bees are kept on the A. & M. College grounds for use in connection with a course in beekeeping given to students in the Agricultural Courses of study, and for demonstration purposes. A more complete apiary, well-equipped with modern tools and implements, is maintained by the present State Entomologist on the Brazos River, about seven miles from the College. This apiary contains at present forty colonies and is used for experimental work exclusively. A small apiary for studying conditions in the southwestern portion of the State has recently been established on the State Experimental Farm at Beeville, Bee County. The results of experiments under way, and reports on the work of foul brood eradication in Texas, will be given in future bulletins of a more or less technical nature.

Nearly every farmer is so situated that a few colonies of bees can be kept to advantage, but many are deterred from taking up the work for lack of sufficient information. Below we give, in as brief a manner as possible, the gist of what everyone should know when taking up beekeeping, either as an incidental line of work or as one of his principal sources of revenue. As the reader progresses in experience and knowledge the facts here given can readily be supplemented by reading any of the many excellent text-books on the subject.

DESIRABILITY OF KEEPING BEES.

In answer to the question: "Will it pay to keep bees?"—everything that can be said must be in the affirmative. There are but few people who do not like honey and its production in sufficient quantities for family use is neither difficult nor expensive. Even if one does not consider the sale of any surplus honey, he will find himself well repaid for investment in a few colonies by the pleasure and satisfaction of having pure and wholesome honey for his own table. An outfit requiring an initial outlay of from $5.00 to $15.00 is sufficient for a beginning in bee culture. In this connection it should be said that only modern frame hives should be used, for there is little but annoyance, discomfort and loss to be derived from keeping bees in box hives or "gums."

Honey is a wholesome food and one that is not without medicinal value. Persons afflicted with chronic constipation often find permanent relief as a result of eating honey daily. Honey is one of the principal ingredients of soothing syrups and cough medicines and most of our elders know full well how to utilize it in making most excellent and efficacious home-made cough syrups. Commercially, honey is also used in the manufacture of various cakes and cookies as it is the only substance which will keep them soft and moist for a long time. Relative to the food value of honey, Dr. C. C. Miller, a prominent beekeeper of Illinois and author of "Fifty Years Among the Bees," says:

"About sixty pounds of sugar on the average is annually consumed by every man, woman and child in the United States. It is only within the last generation that refined sugars have become so low in price that they may be commonly used by the poorest families. Formerly honey was the principal sweet. It would be greatly for the health of the present generation if honey could be at least partially restored to its former place as a common article of diet. The almost universal craving for sweets of some kind shows a need of the system in that direction; but the excessive use of sugar brings in its train a long list of ills. Besides the various disorders of the alimentary canal, that dread scourge, Bright's Disease, is credited with being one of the results of sugar eating."

Prof. A. J. Cook, until recently at the head of the Department of Biology of Pomona College and author of the "Manual of the Apiary," also says:

"There can be no doubt but that in eating honey our digestive machinery is saved work that it would have to perform if we ate cane sugar; and in case it is overworked and feeble, this may be just the respite that will save from a breakdown.

"We all know how children long for candy. This longing voices a need, and is another evidence of the necessity of sugar in our diet. Children should be given all the honey at meal time that they will eat. It is safer; will largely do away with the inordinate longing for candy and other sweets; and in lessening the desire will doubtless diminish the amount of cane sugar eaten.

"Sugar is much used in hot drinks, as in coffee and tea. The substitution of a mild flavored honey in such cases may be a very profitable thing for the health. Indeed, it would be better for the health if the only hot drink were what is called in Germany 'honey-tea'—a cup of hot water with one or two tablespoonfuls of extracted honey."

The importance of bees in fertilizing the blossoms of fruit trees and of other plants, such as clovers, buckwheats, etc., should not be lost sight of. Not infrequently orchards have been found to increase their yields of fruit very perceptibly when bees have been brought into their vicinity. In this connection it should be pointed out that honey bees are not injurious to fruits. They never injure sound fruits of any kind, and it has been repeatedly demonstrated that their mandibles are not powerful enough to pierce or break the skin of any fruit ordinarily grown. When ripe fruits are broken open or injured by other insects, by birds or by accident the bees will often lap up the juices exuding from the wounds. In such cases they are but utilizing what would otherwise go to waste and no greater mistake could be made than that of charging them with being the cause of the injury.

Bees of the proper kind may be kept without danger of people being stung, even in the heart of villages and cities, if reasonable precaution is observed. Bees vary in their dispositions as much as do other domestic animals. One is not likely to be censured for keeping a gentle collie on the ground that some bulldogs have been known to bite. No more should a man be censured for keeping gentle bees in his dooryard because of the fact that some bees are known to be vicious and dangerous. For use in crowded quarters, where many people are about, no bees are better than the Carniolans or a gentle strain of Italians. The Carniolans, particularly, are very gentle and from their behavior would not ordinarily be suspected of owning stings. Such bees as the Cyprians, hybrids between Italian and black bees, or even black bees themselves, should not be kept on city premises or where many people have occasion to come near them.

The profit from keeping bees on a commercial scale is easy to calculate "on paper," but is not always so easy to realize in practice. It not infrequently happens that good colonies, properly cared for, in favorable seasons yield from 40 to 60 pounds of honey, selling at prices varying all the way from seven to fifteen cents per pound. The cash revenue from an apiary under such conditions is of course considerable, but unfavorable seasons, disease, lack of skill or insufficient attention on the part of the beekeeper may reduce the yield to much less than this amount or even wipe it out entirely. Experienced beekeepers have learned that they cannot count on handsome profits every season but find that by judicious management and by caring for their bees in bad as well as in good seasons, they get a good average return from their investments. As to the question of engaging in apiculture on a commercial scale one bit of most excellent advice can be given: Do not enter beekeeping upon a large scale under any circumstances unless you are an experienced beekeeper. Commence with a few colonies and increase in knowledge and experience as the colonies increase. If the colonies fail to increase it is a foregone conclusion that you cannot succeed in the management of a large apiary or apiaries. Many sad disappointments and much lost capital amply attest the fact that this is one pursuit in which the novice cannot expect to succeed on a commercial scale. This, however, should not discourage anyone from finding both pleasure and profit in keeping a few colonies.

HOW TO SECURE BEES.

If the beginner does not happen to have bees in box hives, the best plan for obtaining them will be to purchase from one to five colonies, if they can be secured in his neighborhood. What constitutes a reasonable price for colonies depends upon many factors, chief among them being the extent of the honey crop, the strength of the colonies, amount of honey contained in the hives, the purity of the queens, and the nature and condition of the hive itself. As a general rule, and in most localities, Italian or black bees, in box hives, are worth from $1.00 to $1.50 per colony. Black bees or hybrids, with queens of the same kind, in dovetailed ten-frame hives, are worth probably from $3.50 to $4.00 per colony. Pure Italian bees, in sound dovetailed hives, are worth at least an average price of about $5.00 per colony. These prices may be higher or lower, according to relative supply and demand and according to whether the purchase is made just before the honey season or after it is over.

Colonies in bee-trees, as a usual thing, are worth no more than the labor required for taking them out and installing them in frame hives. At times, however, sufficient honey is secured from such trees to make it worth while to remove the bees from them.

The reader is advised to make his beginning with colonies already in frame hives, as the task of "transferring" colonies from bee-trees or box hives to modern frame hives is not one that is usually enjoyed by the novice. However, instructions for transferring will be found on a subsequent page.

LIFE HISTORY OF THE HONEY BEE.

To be a successful beekeeper, one must have a fair understanding of the domestic life within the hive, and must understand how bees are reared, comb built, honey stored, etc. A knowledge of these things enables the beekeeper to understand what operations may be performed without disturbing the domestic economy of the colony. Nature has endowed the bees with certain definite instincts. Man cannot change these instincts, but he can so work in accord with them that the bees, aided by their master's intelligence, can accomplish far more than they could without it. By understanding the life history and habits of the bees, the beekeeper cannot only increase the honey production, but he can protect the bees from enemies and disease and do it without causing the death of a single bee and without interrupting the regular work of the bees for more than a few minutes at a time.

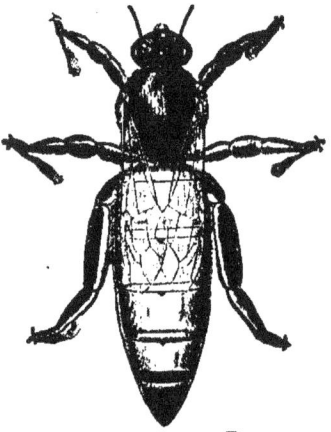

Inside the hive will be found three distinct forms of adult bees: the queen (Fig. 1), the worker (Fig. 2), and the drone (Fig. 3).

Only one queen is normally found in each colony, and her duty is to deposit the eggs from which all bees are hatched. She alone is the egg-layer and is the mother of all bees raised in the colony. Having no other duty to perform, she is specially adapted by nature for her egg-laying duties.

Fig. 1—Queen Bee. Fig. 2—Worker Bee.

She is even fed and groomed by the worker bees, some of which accompany her at practically all times and attend to her wants. From cell to cell she goes, depositing in each a small white egg, cylindrical in form and measuring approximately 1.8 millimetres (about 7-100 of an inch) in length (See Fig. 4).

In appearance, the queen is much longer than the worker bee and is also slightly thicker and broader. However, on account of her lengthened abdomen, she appears much more slender. Her abdomen is not as distinctly banded as that of the worker, and is usually of a uniform tawny or brown color. The thorax is almost devoid of hairs and presents a shiny appearance. A good queen will, during the height of the honey season, deposit as many as 1500 to 2000 eggs per day. Usually she will continue active egg-laying for two years, sometimes three, after which time, if accidental death does not overtake her, the bees see that she is superseded by a young queen.

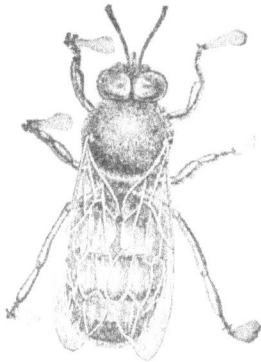

Fig. 3—Drone.

The workers are by far the most abundant individuals of the colony, numbering in strong colonies as many as 40,000. They are the units of organized labor, and to their lot fall all the duties of the hive, barring egg-laying and fertilization of the queen. They build the comb, gather the honey, feed the young bees or larvae and protect the community from robber bees and other enemies. In appearance, the workers differ from the queen in their shorter and smaller bodies and by a marked pubescence upon the thorax. Both the queen and workers are provided with stings, but the queen rarely makes use of hers, even when roughly handled in the fingers or even mashed. It is thought that she uses her sting only in battles with rival queens, a wise provision of Nature whereby this all-important individual shall not endanger her life except in determining the "survival of the fittest." The workers, when angered, use their stings with telling effect, as everyone knows.

The drones are found more abundant at certain seasons than at others. Especially are they plentiful at the approach of the swarming season in spring. Rarely more than two or three hundred are found in a normal colony, and as a usual thing not more than seventy-five to a hundred during summer and autumn. The drones may be readily recognized by their large size, as compared to the workers, and also by the fact that their abdomens are blunt and rounded, instead of sharp. Big and clumsy, very noisy, they perform no part of the daily labor in the hive, for their sole function is to mate with the young queens, which, under ordinary conditions, appear in the colony about once a year, during the swarming season. They gather no honey, do no work, and consume large amounts of the stores gathered by the busy workers. At the approach of winter, or upon a dearth of honey, they are all killed or expelled

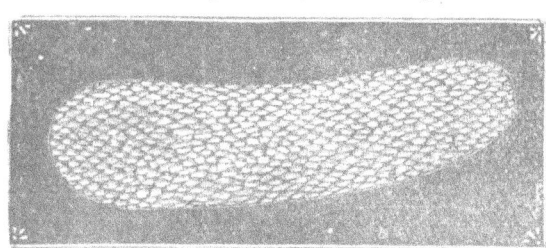

Fig 4—Queen's egg, highly magnified.

from the hive, doubtless as a matter of economy. The drone develops from an unfertilized egg, and as the queen appears to have the power of laying fertilized or unfertilized eggs at will, drones are produced whenever needed, as at the beginning of the next honey flow or at the approach of swarming time.

The egg, already mentioned, when seen under the microscope, presents a beautiful appearance, its surface being covered with a fine network of lines. About three days after it is deposited the egg hatches into a small white larva, or "grub." At or just before the time of hatching it is surrounded by a liquid, milk-like food by the nurse bees. The growth of the larva is very rapid. For about six days it is fed by the worker bees, which are serving duty

Fig. 5—A frame of bees and honey, just from the hive (original).

as nurses, and about ten days after deposition of the egg the larva is sealed over with a thin covering and left to go through its third stage of existence, known as the "pupal stage." It remains sealed over for ten or eleven days, during which time it gradually assumes the appearance of an adult bee. Twenty-one days after deposition of the egg in the case of young workers, and twenty-four days in the case of drones, the bee completes its growth, gnaws open the covering of the cell and emerges a perfect, full-grown bee. Soon it grooms itself, straightening out the small hairs covering its body, and then proceeds to a cell of honey and gets its first square meal as an adult. After the first or second day out of the cell the young worker bee takes up the duties of "nurse," and gives food to the larvae. When the adult bee becomes older it takes up the work of comb building, honey gathering, etc.

The young workers in the hive do not take much part in protecting the colony against intruders. Bees that have just emerged from the brood cells can be handled with impunity. It is the middle-aged and old bees in the hive that make trouble for the operator when the hive is opened or disturbed.

In looking over the combs in a hive the majority of cells will be found of a uniform size. These are the worker cells. In spots, and rarely occupying entire frames, will be noticed cells which are markedly larger than the rest. These are the drone-cells, making up what is termed the "drone-comb." It is in these large cells that the drones are reared. No material difference in the larval history of the drone is noted, as compared to that of the worker, except that twenty-four days instead of twenty-one are required for the attainment of maturity.

The life history of the queen is discussed on a subsequent page under the head of "Swarming."

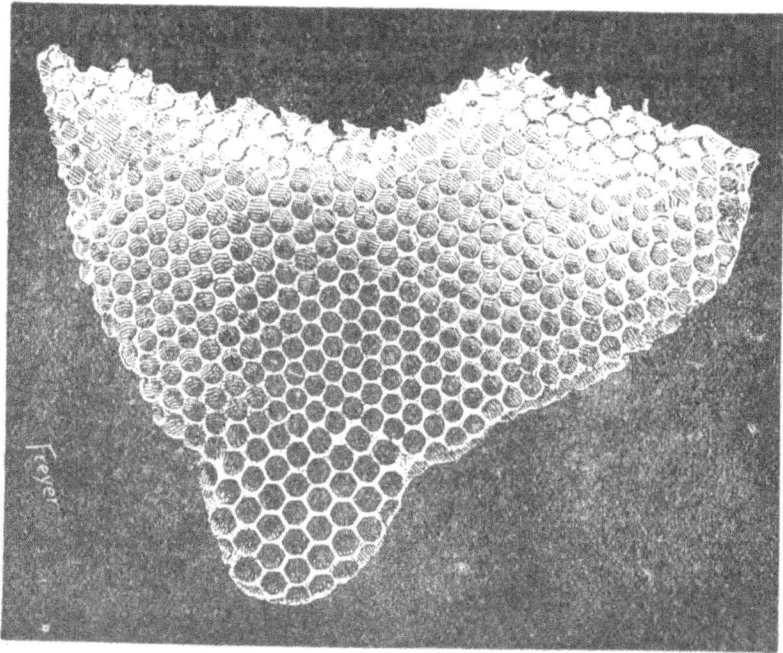

Fig. 6—Piece of comb showing both worker and drone cells, the latter largest and occupying the lower portion (after Phillips, Farmers' Bulletin No. 397, U. S. Dept. of Agr.).

HIVES.

The "hive" is the box or "house" in which the bees are domiciled. That they may be properly observed, that queenlessness may be avoided, and remedied when it does occur, and in order to see that the bees do not run short of stores, as sometimes occurs when the beekeeper has "robbed" the colonies too closely, it is necessary to use a hive from which the frames can be readily removed for examination. We, therefore, have what is designated as the "frame hive." Many different sizes and patterns have been manufactured and a great variety are in use at present, each having its advocates. The hive in most general use, and doubtless better adapted to Texas conditions than any other, as well as being admirably suited for the beginner, is the one known as the ten-frame dovetailed hive. Such a hive, with a "super" or upper story, is shown in Figure 7.

The same style of hive is also made smaller, so that it holds but eight frames. Formerly the eight-frame size was quite popular but of recent years it is going out of use and it seems quite certain that under ordinary conditions a vigorous colony with a prolific queen needs more room in the brood-chamber than is available in the eight-frame hive. The ten-frame hive is at present the one in most extensive use.

The complete hive consists of bottom-board, hive-body, super and cover. The bottom-board has a narrow cleat nailed on top of each side and on one end. The hive-body, consisting of a rectangular, dovetailed box, rests upon the bottom-board and the cleats supporting it leave a narrow entrance for the bees at one end. This is designated as the "front" of the hive, and it constitutes the only pasageway which the bees have for entering or leaving the hive. Inside the hive-body are ten frames, having protruding top-bars, which rest upon tin rabbets within each end of the hive-body. One or all of the ten frames can be removed from the hive at pleasure, and in each one the bees construct a comb.

Fig. 7—Ten-frame, dovetailed hive with super. The larger frame seen on the outside is a "brood-frame," taken from the lower part, or "hive-body." The smaller frame belongs in the super, or upper story. Both frames are seen filled with comb "foundation" (original).

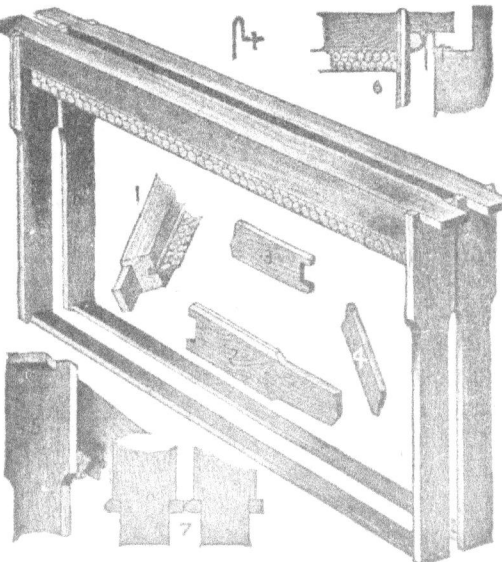

Fig 8—Hoffman self-spacing brood-frames.

As with hives, we have also a variety of frames. One of the most popular and useful styles is that designated as the "Hoffman." By reference to figure 8 it will be seen that one side of the end of each frame has, toward the top, a V-shaped projection, while the opposite side has a flat surface. When the frames are placed side by side in the hive, these V-shaped projections keep the frames a "bee-space," or the proper distance apart. This is an important feature, for when the frames are placed too close together insufficient room is left for the bees to cluster above the brood and to pass to and fro in their work. On the other hand, if the frames are too far apart, the bees will attempt to fill up the space by either building additional parrallel combs outside the frames or "brace-combs" from one frame to another, both of which are an aggravation to the beekeeper. By using the Hoffman, or a similar frame, the beginner avoids this difficulty, for these frames are "self-spacing." Frames which are not self-spacing are now rarely used by progressive beekeepers, owing to the loss of time involved in spacing them by hand each time the colony is examined. One of the marked improvements made in frames in recent years is the metal-spaced frame. In principle, this frame is not unlike the Hoffman, but, instead of having V-shaped projections on the end pieces, it has a stamped metal band running over the ends of the top and down along the ends. This metal band has on it projections which keep the frames the proper distance apart. Being tacked securely to both the top-bar and the ends, this band gives the frame additional strength. The bees are not able to fasten together, by means of bee-glue, or propolis, the metal-spaced frames as they do the all-wood frames. As a great amount of propolis is gathered by the bees in Central and East Texas, the metal-spaced frame is preferable to all others in these localities. These frames are slightly more expensive than the all-wood frames, but, as they outlast the latter, they are much the cheapest in the end.

Before the frames are placed in the hive, there should be placed in each sheet of comb "foundation." This is made by machinery and consists of a sheet of pure beeswax having impressed in it, on both sides, the bottoms of the cells, as well as portions of the side walls. As its name implies, it is the foundation upon which the bees build the comb. By its use much time and labor is saved the bees, to say nothing of the fact that it insures straight combs and reduces the amount of drone-comb to a minimum. This foundation is fastened into the frame with a double groove and wedge (Fig. 9) by means of melted wax or with a foundation roller. After being inserted in the frame, the roundation should have imbedded in it three or four small tinned wires, which have previously been tightly stretched through the frame from end to end. The imbedding of the wires in the foundation is done with a "spur wire-imbedder," which is merely a small steel tracing wheel with alternating teeth. These wires serve to hold the comb firmly when built out and to prevent its breaking or sagging when shipped or handled. Under no circumstances should this wiring be omitted, for during extremely hot summer weather the combs become very soft and easily broken. They have even been known to melt under the sun's heat when not properly shaded during the hot months. Upon the hive-body is

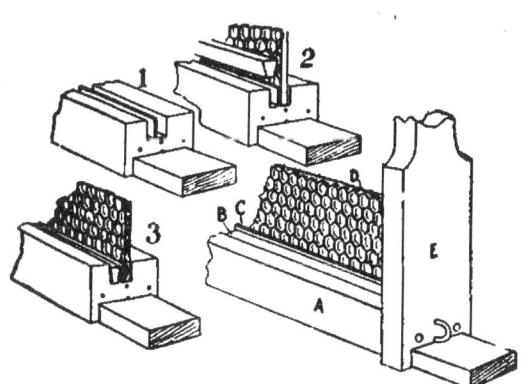

Fig. 9—Method of fastening foundation in frame with a small wooden wedge (courtesy A. I. Root Co.).

placed the cover. Covers are also made in various designs, but a good cover should be water-tight and the boards should be prevented from warping by being inserted in grooved end-pieces. During the summer months, if the hives are not located in a shady place, a "shade-board" must also be placed on top of the hives to protect them from the sun. The shade-boards can be made of rough lumber, old boxes or of anything that "comes handy."

When surplus honey is being gathered by the bees a super, or upper story, is placed upon the hive-body and the cover on top of the super. At times as many as three or four supers are necessary on the hive at one time, to afford the bees sufficient room in which to work and store the honey. (Read the paragraph on "Supers," page 26.)

OTHER EQUIPMENT.

In addition to hives and bees, it is imperative that the beekeeper have a strong veil and good smoker. Veils of various styles and prices may be secured from dealers in bee supplies. One of the good ones on the market is made from black cotton tulle, with a face of fine silk tulle. The silk tulle does not interfere materially with the sight. We have found, however, that cockroaches and crickets are seemingly very fond of the tulle, and sooner or later they find opportunity to detsroy the veil. We have, therefore, adopted a veil made principally of wire cloth, with hat piece and neck piece of heavy muslin. This veil, besides being insect-proof, admits more air to the face of the wearer than do the cotton veils. As the wire cloth can not touch the face at any point, it is impossible for the bees to sting through it. A good veil may be made at home of either tulle or mosquito netting. A band of rubber should be inserted in the upper end of the veil to hold it snugly around the hat brim, and the veil itself should be of ample length so it can be tucked beneath the coat or vest.

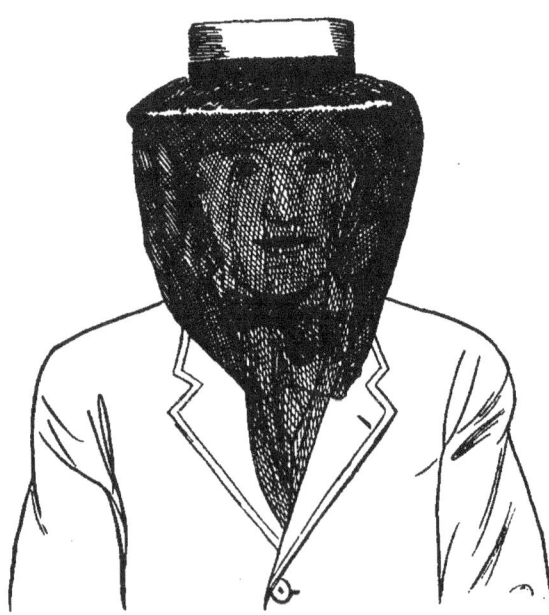

Fig. 10—Bee veil of tulle (after Phillips, Farmers' Bulletin No. 397, U. S. Dept. of Agriculture).

It is also well to be provided with a pair of heavy cloth or medium weight leather gloves, preferably with long sleeves. Cotton gloves, either with or without fingers, are supplied by manufacturers of beekeepers' supplies. In warm weather the beekeeper will soon prefer working without gloves, but it is well to have them constantly at hand for use with cross colonies or in case a "spree" of stinging is started by robbers. The use of the veil should not be dispensed with, even with gentle bees, for unexpected attacks will sometimes occur in spite of all precautions. The bees can be prevented from crawling up the trousers legs by tucking the latter in the stockings, by wearing leggins or by securing the trousers with bicycle pants-guards.

A good smoker is necessary. Smokers may be purchased at prices varying from thirty or forty cents to several dollars. For fuel in the smoker almost any dry material will answer. Dry rotten wood, fine shavings, old rags, and oily waste, such as is obtainable from machine shops, are excellent fuels. Sulphur should never be used in a smoker and tobacco but rarely. We have at times addded a little tobacco to the fuel in the smoker in order to subdue bees that were exceptionally vicious and that had to be manipulated or inspected at the time. The tobacco smoke stupefies the bees and if used too heavily will even kill them.

An outdoor box or other receptacle should be provided in which to keep the smoker when not in use. Sparks often remain in the smoker for hours after it has been emptied, hence it should not be kept in a building. Our plan is to nail a good water-proof box to a tree in the apiary, the box having a good door and lock and in this the smoker is kept at all times.

After considerable experimenting we have hit upon a method of lighting the smoker which rarely fails. A small handful of excelsior is pressed into firm lump, held over the smoker barrel and a match applied to its under side. As soon as it catches it is pushed down in the barrel and dry, rotten wood placed on top of it, the bellows being worked in the meantime. By this plan it is easy to light the smoker in a high wind, or even during a rain, provided only that the excelsior and wood are dry.

HOW TO OPEN A HIVE.

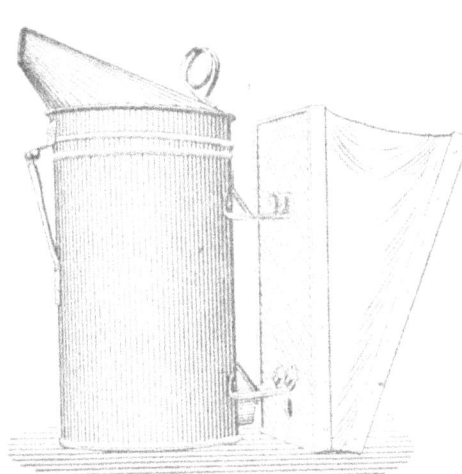

Fig. 11—Smoker (after Phillips, Farmers' Bulletin No. 397, U. S. Dept. of Agriculture).

Having adjusted the veil, and having the smoker well going, blow one or two puffs into the entrance of the hive to be opened. Do this with a strong closing of the smoker bellows, so as to drive the smoke thoroughly into every part of the hive. This does not mean that the bees should be deluged with smoke. All that is required is that each bee in the hive should get a whiff, however light it may be. Next grasp the hive-cover by one end and gently raise it, blowing a puff or two of smoke into the crevice thus made. In the majority of cases this will be sufficient, and further smoke will be unnecessary. In case, while working with them, they fly out and attempt to sting, a little more smoke may be blown into the hive from above. On general principles, no more smoke should be used than is necessary to prevent them from stinging. Smoke demoralizes the bees, and as considerable time is required for them to recover from a severe smoking, much time, and sometimes honey, is lost by them. As soon as smoke enters the hive, and the bees are disturbed, the majority of the workers go to the unsealed honey and proceed to fill up, presumably on the supposition that their home is about to be destroyed and in being compelled to leave they purpose to take with them as much of their treasure as possible. Why smoke should have such an effect upon the bees is often a source of wonder to many people. The probable explanation is simple. All wild creatures are afraid of fire and the bees, not unlike other creatures, have learned by generations of experience that fire is a force which they can not hope to combat successfully. As smoke is the forerunner of the fire they doubtless conclude that it is better to load up with their household supplies of honey and prepare to vacate than to attempt opposition to the smoker and its operator. When a bee is well gorged with honey it will show no disposition to sting, and where exceptionally cross colonies are to be managed they will be found more docile if a puff of smoke is blown into the entrance about five minutes before opening the hive. This will give the bees time to gorge themselves with honey and they will not be as pugnacious as they otherwise would be. In opening the hive, if the cover sticks, do not jerk it off, but use a small screwdriver, pocketknife, hive tool (Fig. 12) or other instrument to gently pry it. Having removed the cover, jar the bees off in front of the hive, turn the cover on edge and sit down on it. Now gently push the frames apart, or, if stuck tightly, pry them apart with chisel or knife and lift out the one that is to be examined.

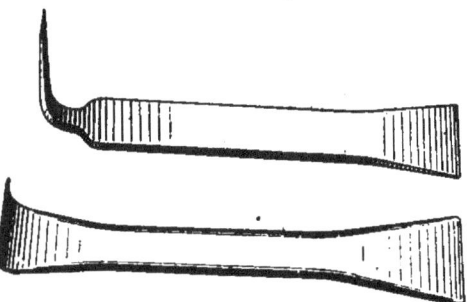

Fig. 12—Hive tools (after Phillips, Farmers' Bulletin No. 397, U. S. Dept. of Agriculture).

It will be noticed that where the cover joins the hive, and where the frames touch each other at the ends of the hive, is found a grayish-yellow substance of a very adhesive nature. This is "propolis," or bee-glue, which is gathered from various plants and flowers, and is used as cement for making secure all parts and for sealing up all openings and crevices in the hive, except, of course, the entrance. It is this material that sometimes causes the cover to adhere tightly and the frames to stick together. Lift the frame by the ends of the top-bars and keep the entire frame vertical. When it is desired to examine the reverse side of the frame, turn it up on end, without changing the hold, always keeping the comb in a vertical plane, and revolve it through a half-circle, using the top-bar as a pivot. The other side will now be toward the operator, and can be examined at leisure. This method of handling is to prevent any possibility of the comb breaking or sagging, and is especially necessary in hot weather and when combs are not wired. It is based upon the very simple principle that the comb hanging directly down from the top-bar, or supported by the top-bar when reversed, is not nearly so liable to sag as when the comb is held in a horizontal plane supported by the top-bar at one edge only. If the colony is in prime condition, plenty of bees will be found clustered on all combs, the two frames next the outside of the hive will be found filled with honey, much of it sealed, and the ends and about one or two inches along the tops of all other combs will also contain honey. In the central frames of the hive will be found sealed brood, which, by its even surface and brown color, is readily distinguished from the sealed honey, unsealed larvae and eggs, each of the latter being attached by its end to the bottom of the cell. This part of the hive containing eggs and larvae is designated as the "brood-nest" of the queen, and in it, or near its outer edge, she will usually be found. At times she appears to take little excursions to the most remote parts of the hive, probably by way of exploration, and may be found in a remote corner. Especially if much smoke has been used in opening the hive, the queen will become alarmed and make every effort to elude the operator by running from comb to comb, dodging around corners of the frames, and crowding under thick clusters of bees. A little practice will enable the beekeepr to locate her easily. The brood-nest may cover the greater part of seven or eight, or even ten frames during the height of the honey season, and from this number of frames on down to one or two in the case of a dearth of honey. During the winter season a relatively small amount of brood will be found.

After examining the colony, the frames should be replaced in the same order in which they were prior to opening, in order that the brood-nest of the queen may not be disorganized. Having the frames all replaced, push them over against one side of the hive, as close as they will go, and as close to each other as the V-shaped projections will allow. They will then be correctly "spaced," without further attention. This will, of course, apply only to the Hoffman or other self-spacing frame. In some localities the "all-wood" frames are used extensively. These latter are not provided with any self-spacing device and must be placed the proper distance (one-fourth inch) apart by the operator. Their only advantage lies in the fact that they are a trifle cheaper than the Hoffman frame, but the time lost in spacing them correctly is worth more than the additional cost of the Hoffman or the metal-spaced frames. The hive, after each examination, should be carefully closed, pains being taken to see that no cracks or openings are left between the hive and cover through which robbers might enter.

It is well to keep a stone, brick or other weight upon the covers at all times to prevent the hives being overturned by hard winds. In summer, shade-boards should be provided, as above suggested.

LOCATION OF HIVES AND APIARY.

Perhaps the best location for an apiary is in a shady grove of deciduous trees, where there is ample ventilation, and where the limbs are sufficiently high so as not to interfere with the flight of the bees or with the work of the beekeeper. Places which are densely shaded are to be avoided, for in such places dampness often causes the unsealed honey and the pollen stored in the hive to ferment or mould, to the detriment of the bees.

Each hive should be slightly raised from the ground, either by means of bricks or by specially constructed hive-stands. The hive should be slightly raised—one or two inches—at its rear end, so that rain water beating in the entrance will quickly drain off.

All grass and weeds around the hive should be kept cut down, so as to afford the bees free passageway to and from the hive.

The direction in which the entrances face appears to be of little consequence, except that they should never be toward the north or northwest. Many beekeepers prefer to have the hives facing east or southeast, on the theory that the early morning sun will warm the hives and induce the bees to begin work early.

In hilly sections, the apiary should always be located on the east or south slope.

In the location of "out-apiaries," or "out-yards," as they are more commonly called, a number of important points must be considered in choosing the location. An out-apiary is usually situated several miles from the beekeeper's home and in attending to it he must travel back and forth a good many times a year. The first consideration is, of course, the honey-bearing flora of the neighborhood, for no apiary can be successful unless the forage for the bees is good. Other things being equal the out-apiary should be so located that it can be reached over good roads, rather than bad ones, as this means a big economy in time and in the cost of transporting supplies and honey to and from the apiary. A water supply should also be within reach of the apiary in order that the bees may get water at all times of the year. Temporary supplies of water, such as shallow ponds and "tanks" should not be depended on for they will fail during severe drouth, just at the time when the bees will need water the most. In short, common sense and good judgment should always govern the location of both hives and apiaries.

EQUIPMENT FOR EXTENSIVE WORK.

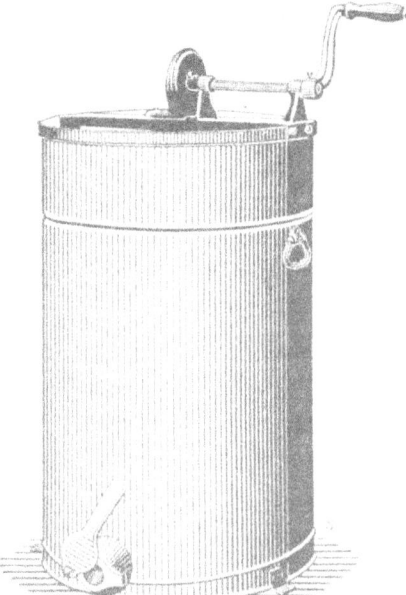

Fig. 13—Two-frame honey extractor (after Phillips, Farmers' Bulletin No. 397, U. S. Dept. of Agriculture).

While the outfit described above will suffice for the small beekeeper who is producing honey for his own use, and only a limited amount for sale, he will find that there are a number of implements very convenient, and in case he intends enlarging his apiary ultimately, implements that are necessary in saving time and putting his products into marketable shape. A few of these will here be mentioned. A honey extractor is one of the most useful and labor-saving devices yet invented for the beekeeper. Its purpose is to extract the liquid honey from the combs, so that the latter may be returned to the bees to be again filled by them. This saves much time during the height of a honey flow. As it requires several pounds of honey to make a single pound of comb, it will be seen that much honey is saved by thus using the same combs over and over again. The honey extractor is constructed on the same principle as is the cream separator, *i. e.*, centrifugal force. The extractor consists of a galvanized iron can, having within it two, four or more pockets, each of which will receive a frame or comb of honey. These pockets are mounted upon a frame pivoted at top and bottom, so that it can be revolved rapidly by means of attached gearing and handle. The rapid motion throws out the honey from the outer cells of the comb, whence it runs down the sides of the extractor and is drawn off through a gate below. The frames are now reversed or turned with the other face to the outside of the extractor, and the frame is again revolved, throwing out the remaining honey. The combs are now ready to be returned to the hive. Honey extractors are manufactured in various styles and sizes adapted to more or less extensive work. A well-built extractor, capable of holding two combs at a time, such as is shown in figure 13, is amply large enough for an apiary of fifty colonies or less. Along with the extractor a honey knife, or uncapping knife (See Fig. 14), will be needed for cutting comb, cleaning, and especially for shaving the cappings off the sealed honey before it is placed in the extractor. The uncapping knife should be kept *very* sharp, and it will be found to work best if immersed from time to time, during uncapping, in hot water kept near the uncapping can or extractor.

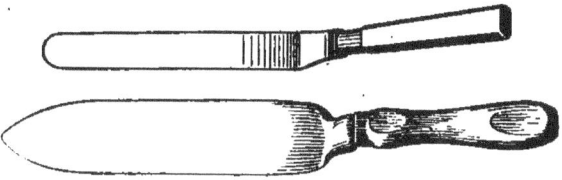

Fig. 14.—Uncapping knives (after Phillips, Farmers' Bulletin No. 397, U. S. Dept. of Agriculture).

While the two-frame extractor is amply large enough to meet the needs of the average beekeeper, extractors are also made in larger sizes, holding four, six or eight frames at a time. The six and eight-frame sizes are rather heavy for operation by hand and they are usually run by a small gasoline engine of one or two horse-power. In very large extracting plants the engine also operates a small honey pump which receives the honey as it comes from the extractor and transfers it to a large storage tank from which it is later drawn and placed in cans or barrels for shipment.

An uncapping can or uncapping box is quite essential when the honey crop reaches several hundred pounds or more. This is either a can or box of convenient size having a slatted or screened bottom to catch the cappings as they fall from the knife and having means for letting the honey drain from the cappings to be drawn off in a separate vessel. The combs, while having the cappings shaved off, are allowed to rest on a simple framework above the uncapping can. Descriptions of the various styles of uncapping cans and boxes will be found in the catalogue of any manufacturer of beekeepers' supplies. In recent years uncapping cans have been devised which melt the cappings and automatically separate the wax and the honey. The cappings should never be thrown away for they are of nearly pure wax and the beeswax obtained from melting them is the very choicest and prettiest obtainable. The cappings will furnish approximately one pound of wax for each one hundred pounds of honey extracted. In a recent experiment in which 1500 pounds of honey were extracted we carefully saved all cappings and they yielded, when melted in the solar wax extractor mentioned below, a total of 15¼ pounds of choice yellow beeswax.

Next in importance comes the solar wax extractor for reducing combs, cappings and scrappings to wax. This consists of a box covered with glass, and having inside it a metal tray painted black, into which the combs are dropped. At the lower end of this tray are one or more tin vessels for catching wax and honey. When placed in the sun, the heat of the latter melts the comb and allows all honey, and the greater part of the wax contained, to run down into the vessels below, from which they can be taken at leisure. After being allowed to cool, the wax in the pans will solidify and can be taken out in cakes. This wax extractor should be kept in the apiary, exposed to the direct rays of the sun, and into it should be thrown all old or broken combs, scrappings of wax, etc. The wax obtained in this way will in a short time pay for the original cost of the device.

When the apiary reaches considerable size an extracting house, or "honey house" in which to keep extra supplies, tools, etc., and in which to extract and pack honey, becomes desirable. If the apiary is located at one's home it is often practical to use a room in some regular farm building for this purpose, but when the apiary is located at a distance, a small building may be erected to advantage. A small shed, with clay or wood floor, will answer very well for extracting purposes if made and screened so that bees can not get into it during extracting time.

A number of Texas beekeepers use tents in which to extract, moving these tents from one apiary to another as needed. Usually such tents are fitted with wire cloth in the gable ends so as to give better ventilation. Still other beekeepers construct large wirecloth cages, large enough to hold the uncapping can and extractor, and still others use an extracting cage mounted upon a wagon. The uncapping, extracting, etc., is done inside the wagon cage, which is covered with wire cloth and has a screen door and as soon as the extracting is finished at one apiary, the wagon and outfit is driven to the next one. The individual beekeeper will have little difficulty in fitting up an extracting outfit to meet his particular needs.

The bee-escape, shown in figure 15, should be placed at the tops of screen doors and windows in the extracting room to permit the escape of bees carried in with the combs. These bee-escapes can also be placed in honey-boards (Fig. 17) in such a manner that the latter, when inserted between super and hive-body, will permit the bees to leave the super but not re-enter it. In this way all bees may be cleared from the super before the hive is opened for the removal of surplus honey.

Queen-excluding honey-boards (Fig. 16) are also useful where honey is produced upon a considerable scale. These boards, made

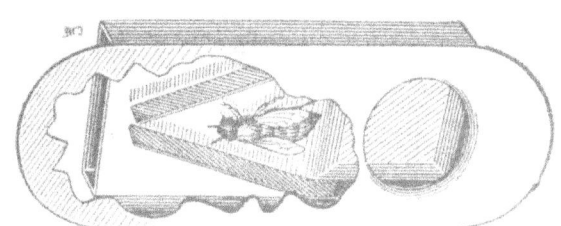

Fig. 15—Porter bee escape (after Phillips, Farmers' Bulletin No. 397, U. S. Dept. of Agriculture).

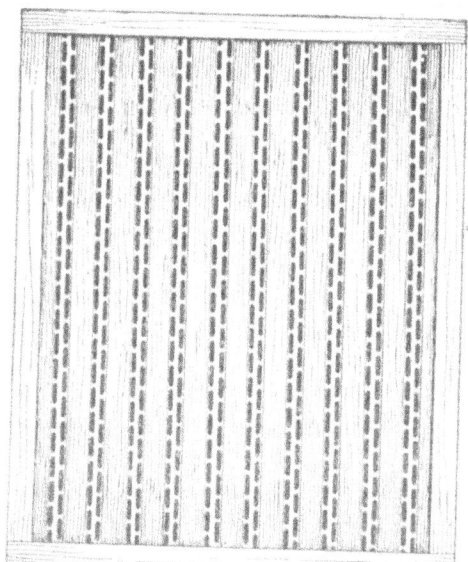

Fig. 16—Queen excluding honey board.

of wood and perforated zinc, have openings which permit the passage of the worker bees, but which are too small to admit the queen. By placing them between hive-body and super the queen is prevented from depositing eggs in the supers, to the disfigurement and impairment of the surplus honey.

TRANSFERRING.

There are several methods of transferring bees from box hives into modern frame hives, or "patent gums," as they are sometimes called. The method selected will depend largely on the preference of the individual beekeeper. The "A. B. C. of Bee Culture," a well-known book on beekeeping, gives the following directions for transferring:

"We will assume that your hive, or hives, having been received in the flat, are put together and painted, and contain frames of wired foundation ready for the bees. Light your smoker and put on your bee veil. Move the old hive back four or five feet, and put the new hive in its place. Prepare a small box about eight inches deep, and one side open, that will just cover (not slip over) the bottom of the box-hive. Turn it upside down; set the hiving box over it, and then drum on the sides of the hive with a couple of sticks until about two-thirds of the bees pass up into the box. Gently lift off the box containing the bees, and dump it in front of the entrance to the new hive. Make sure the queen is among them by watching for her as she passes with the rest into the entrance. If you do not discover her, look inside the hive. If you still fail to find her, drum out bees from the old hive again until you do get her, for, to make the plan a success, she must be in the new hive.

"Return to the box-hive and turn it right side up and set it down a couple of feet back of the new one, with its entrance turned at right angles. You now have in the hive about one-third of the original colony, the combs and all the brood. Allow the old hive to stand for at least twenty-one days, at the end of which time the brood will be hatched out, with the exception of a little drone-brood, which will be of no value. Turn the hive upside down and drum the bees out again into the hiving box, after which dump it in front of the new hive as before. If the queen in the new hive is the one you wish to keep, put an entrance guard over the entrance to catch the young queen hatching in the meantime in the old hive, for she would go in and one or the other would be destroyed. If there is no choice of queens, let the second drive of bees go in and the queens will fight it out. Your job of transferring is now completed, and all you have on hand is an old box-hive containing a lot of crooked combs, with perhaps a little honey and drone-brood on it. The honey can be extracted, or used for chunk honey on the table, if fit for use."

The writer employs a method somewhat different from the above, involving, perhaps, a little more work, and resulting in more stings, but offering the advantage of getting the transferring work over in a short time, and without any further attention being required. The writer's method is as follows:

Bee veil, smoker, new hive, frames, chisel, hammer and a long knife (such as a butcher knife) are provided. Plenty of smoke is first blown into the entrance of the box hive. The latter is then removed from its stand, and the new hive placed in its stead, with entrance facing in the same direction. The top or side of the box hive is then pried off with the chisel, exposing the combs on the inside. Plenty of smoke is used to keep the bees under control and to drive them off of the combs as the latter are removed. The combs are cut out one at a time, and the bees brushed off of them onto the ground in front of the new hive. For this purpose a bee brush (Fig. 18) is very handy.

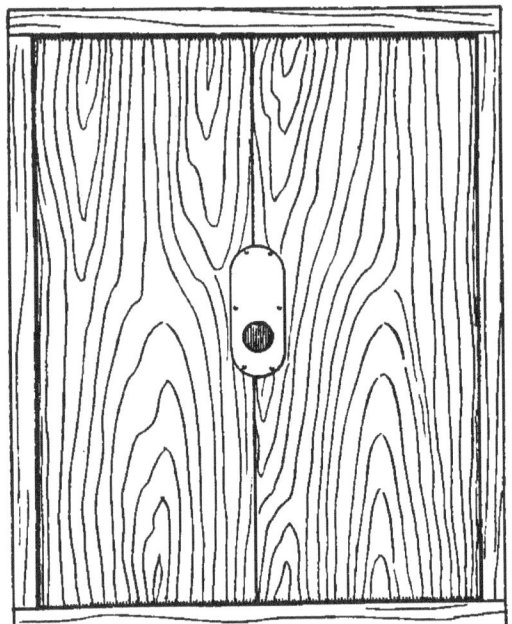

Fig. 17.—Honey-board with bee-escape (after Phillips, Farmers' Bulletin No. 397, U. S. Dept. of Agriculture).

All parts of comb containing either sealed or open honey are cut out and placed in a pan or bucket to be fed back to the bees at some future time. If desired, the sealed honey may be kept for table use. The combs containing brood are cut to fit into the frames of the new hive. These pieces of brood-comb are then fastened into the frames by winding white, soft, cotton cord around the frame several times, from top to bottom, so that the comb is held firmly in place. These frames of brood are then placed in the new hive, and the bees immediately begin clustering on them. In placing this comb in the frame, care should be taken to have the same edge of the comb upwards as was upwards in the box hive. After the honey and brood has been disposed of as above, any bees remaining are shaken down in front of the new hive and allowed to go in.

Usually, enough brood-comb is obtained from a box hive to fill three or four frames in the new hive. The balance of the frames in the new hive should have sheets of foundation (preferably full-sized sheets) placed in them, with the frames wired, and these frames placed in the new hive along with the brood-comb.

If it is desired to feed the honey back to the bees in the new hive— and this is necessary if the transferring is done at a season when the bees are not gathering honey from the fields—this can be done readily by placing an extra super (without any frames in it) on the hive, this super separated from the brood-chamber by a honey-board or super-cover with a half-inch hole in it. A plate or tray of the honey is then placed in the super and the bees, coming up through the hole from the

Fig. 18—Bee brush (after Phillips, Farmers' Bulletin No. 397, U. S. Dept. of Agriculture).

main part of the hive, take it and carry it down into the brood chamber. As rapidly as the plate or "feeder," is empties by the bees, it can be refilled and the feeding thus continued until the colony has stored up enough of the honey to last until the next honey flow or over winter.

About a week or ten days after the date of transferring, the hive should be opened, and if the brood-comb has been securely fastend to the top and ends of the frames the cotton cords should be removed. These should be left on the frame, however, until the comb is securely fastened in place.

Transferring colonies from bee-trees is not particularly different from the process described above, except that one must first cut down the tree and then get at the hollow containing the bees by a combination of sawing and chopping until the combs are made accessible. They are then cut out and secured in the frame hive as above described. It is a good idea to leave the new hive for a few days at the point where the tree was cut open. This not only gives the bees time to fasten the combs in the frames before they are moved, but allows all the bees to find the hive and get accustomed to going in and out of it.

ROBBING.

If bits of honey be dropped about the apiary, or left where bees can obtain access to them, this will start what is termed "robbing." The bees finding this honey will gather it up and carry it away to their respective hives. When this supply is exhausted they will greedily search for more, and if none is to be found will attack some adjacent hive. If the colony in the latter is weak, the invaders will conquer, kill a majority of the rightful occupants, destroy the brood and carry off the honey. If the attacked colony is strong a pitched battle ensues, the result of which will be hundreds of dead bees on both sides, even though the robbers be repulsed. Once in the habit of robbing, this habit is liable to be kept up for days or even weeks. Whenever a hive is opened the robbers are on hand and immediately plunge in. For this reason no honey should ever be left exposd, and during a dearth of honey hives should not be kept open longer than is absolutely necessary.

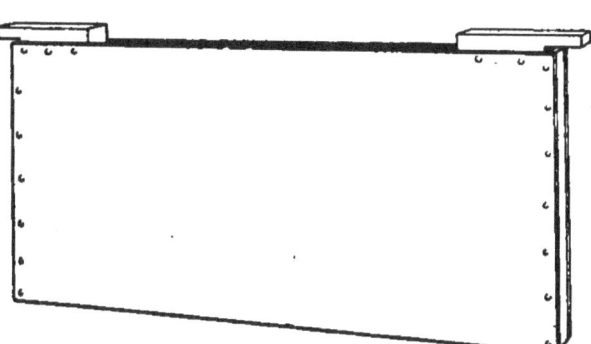

Fig. 19—Division-board feeder (after Phillips, Farmers' Bulletin No. 397, U. S. Dept. of Agr.).

When robbing has once started, the entrances of all adjacent hives, and especially of the hives being attacked, should be closed down to a small aperture. Grass, weeds or Spanish moss thrown over the entrance of the attacked hive will also assist its inmates in repelling the robbers. Sprinkling the front of the hive being robbed with water and carbolic acid is also effective in driving away the robbers.

WINTERING.

In northern states beekeepers find it necessary to place the colonies of bees in cellars during the winter in order to protect them from severe cold.

In the mild climate of Texas such pains are unnecessary, the only precautions necessary being to see that the colonies are supplied with plenty of honey for winter use and that the entrances are not too large. In fact, the latter should be narrowed down severely during the coldest weather.

FEEDING.

If the colonies become weakened from any cause, and there is, coincidentally, a scarcity of honey for them in the fields, it will be found necessary to feed them. Especially must the colonies be fed if they run out of honey during the winter months.

There are a number of devices for feeding the bees, one of which is shown in figure 19. The division-board feeder is inserted in the hive-body in the place of one or two brood-frames and the honey or syrup poured into it from above.

We have often used ordinary dinner plates for feeders, placing a few sticks in them to keep the bees from drowning. The plate is placed in an empty super above the brood-nest and separated from the latter by a super-cover having a half-inch hole in it.

Feed should be made from one part granulated sugar and two parts of water. The syrup should *never* be *boiled*, but if one will heat the water to the boiling point before making the syrup, the sugar will dissolve very quickly indeed. The hot water should be placed in a bucket or tub and the sugar added slowly, the water being stirred vigorously during the operation. The bees should be fed each evening at or a short time before sunset. In feeding take off the hive-cover, insert a funnel into the opening in top of the feeder and pour in the syrup. When feeding to stimulate brood-rearing about one pint of the syrup should be fed daily, and the amount gradually increased as the colony increases in bees and brood. When feeding for stores alone, the syrup should be made of one part sugar to one part water, and can be fed in larger quantities; in fact, as rapidly as the bees will remove it and store it in the combs.

Molasses, sorghum syrup, glucose and the like are not suitable for feeding to the bees, owing to the substances they contain other than sugar. No more economical feed than granulated sugar, or pure honey, is likely to be found.

When honey is used for feeding, the beekeeper should make certain that the honey has not been produced by a diseasd colony.

HONEY FLOWS.

The period during which the bees are gathering honey in considerable amount is designated by beekeepers as the "honey flow." Honey flows may be "light" or "heavy," according to the amount of nectar available in the open flowers or nectaries. The honey flow varies greatly in different localities and in different seasons. Localities which afford at least one heavy honey flow each year are, as a rule, adapted to beekeeping upon a rather extensive scale. On the other hand, localities in which the honey flow is irregular or is uniformly light are adapted to beekeeping upon a small scale only. In such localities the keeping of a few colonies to furnish honey for home consumption will be found most profitable. There are few, if any, localities in Texas that will not support at least a dozen colonies in each location.

The honey flow depends of course upon the number and abundance of nectar-yielding plants, and localities only a few miles apart may present great extremes in this regard. Thus at College Station, in the post oak uplands, the honey flow is comparatively light and no surplus is ordinarily produced except during the blooming of the horsemint in April and May. In some seasons the bees do not gather more than enough honey for their own needs and feeding must sometimes be resorted to in order to carry the colonies over winter. Beekeeping on a commercial scale in such a locality is quite out of the question. On the Brazos river, however, not more than seven miles away, a good honey flow can be counted on and commercial apiaries are a success. In the latter locality a light honey flow is obtained during March and part of April from fruit trees, wild plum, haws, youpon and a number of forest trees. This is followed during May and June by a usually heavy honey flow from horsemint, and this in turn by honey from the cotton grown extensively in the lowlands along the Brazos river.

In the semi-arid sections of Southwestern Texas the honey flows are usually of short duration, occurring during the spring months. As a usual thing the honey flow in such regions is very heavy while it lasts, enabling the bees to store honey very rapidly. The main honey-yielding vegetation of the semi-arid regions is very different from that of Central and East Texas, being composed of a variety of shrubs and small trees known collectively as the "chaparral." Among the most important honey-producing plants in the chaparral are the huajillo[1], cat's-claw[2], huisache[3], eysenhardtia[4] and mesquite[5]. In general, also, it may be said that regions of considerable annual rainfall, such as those of East Texas, usually yield honey of a dark color in abundance, though often of excellent flavor. The semi-arid regions, on the contrary, invariably produce honey very light in color, of excellent flavor and commanding a fancy market price.

The individual beekeeper must become acquainted with the nature of the honey flow in his own locality, and must determine to what extent it is constant from year to year before he can gauge the extent of his apicultural operations.

For the best results in the production of surplus honey it is necessary that the colonies be very strong and have plenty of brood when the main honey flow arrives. To bring about this condition some bee-

---

1 *Havardia brevifolia* Small
2 *Acacia Wrightii* Bentham and *Acacia Greggii* A. Gray.
3 *Vachellia Farnesiana* Wright and Arnott.
4 *Eysenhardtia orthocarpa* S. Watson.
5 *Prosopis glandulosa* Torrey and *Strombocarpa odorata* A. Gray.

keepers induce brood-rearing for two or three weeks before the honey flow by stimulative feeding, that is, by feeding a small amount of honey or syrup daily. When there is a light, steady honey flow preceding the main flow, so that the bees are rearing brood steadily and in goodly amount there is no need of stimulative feeding. In fact, feeding under such conditions is very likely to induce excessive swarming, a thing to be avoided as far as possible. On the other hand, if there is no nectar available for some time preceding the main honey flow, feeding to keep up the rate of brood-rearing will be found profitable. Unfavorable weather conditions preceding the main honey flow may also result in reduced brood-rearing and this can be offset advantageously by feeding.

CULTIVATED HONEY PLANTS.

Many people are under the impression that handsome profits are to be derived from growing honey plants especially for the bees. After many experiments and much observation it must be said that such a plan is, in the great majority of cases, unsuccessful. During the past nine years the State Entomologist's Department at College Station has tested all cultivated honey plants available, such as buckwheat, trefoil, *Euphorbias*, rape, soja beans, cowpeas, velvet beans, vetches, Australian salt bush, borage, mustard, mignonette, privet, various clovers, etc. While these plants yield honey in fair quantity the expense of their cultivation has always exceeded the value of the honey produced, and it is a safe conclusion that the cultivation of crops for honey alone is not profitable in this State.

There are many cases, however, in which a crop that is of value for forage or seed will also yield considerable honey, the latter in such a case being clear profit. Thus in sections adapted to its growth alfalfa is a paying crop on account of its value as hay and a not inconsiderable amount of honey is secured from it. The same remark applies to white clover, which makes excellent pasturage and is a good yielder of nectar. Buckwheat can frequently be profitably grown for seed and considerable honey is secured from it.

On most premises trees and shrubs which yield nectar during their blooming periods can be utilized for ornament. Well placed hedges of California privet add to the beauty of farm premises and the early blooms assist very materially in rearing of the brood prior to the main honey flow. Umbrella china, so frequently planted about farm yards, affords a short but early supply of honey for brood-rearing. Fruit trees furnish a goodly supply of honey very early in the spring, and fruit trees of all kinds will be found to bear more abundantly when honey bees are present to cross-pollinate the blossoms.

SUPERS.

As soon as the bees have filled the hive-body with brood and honey a super, or upper story, should be placed on the hive. Supers are of various kinds, according to the form in which the surplus honey is desired.

If "section honey," that is, honey stored in small frames containing a pound each—ready for market—is desired, a section super must be used. Several styles of sections are obtainable, a full description of which will be found in any bee supply catalog.

The best results in producing section honey are obtained in localities where the honey flow is heavy and fast and where there is but little propolis. When the latter is abundant the bees stick the sections together with it and it is very difficult to have the sections free of propolis stain when they are finished. The production of section honey has not become popular in Texas.

If extracted, or "separatored" honey is to be produced an "extracting super" should be used. This may be either a full depth hive-body, identical with the one used for brood-chamber, or it may be a "shallow extracting super" (see Fig. 7) with frames about half the depth of those in the hive proper. When the full depth super is used the frames should all be wired. With the shallow frames this is not imperative but when extracted honey only is to be produced, it is well to wire these frames also.

Perhaps the most marketable form of honey in the Southwest is that known as "bulk comb," or chunk honey. This is sealed comb honey packed in glass jars, or in tin pails or cans, and having the interstices filled with extracted honey. Many customers have an apathy to extracted honey alone, and many others hold an erroneous idea that honey can not be pure unless they see a piece of comb in it. It is more profitable for the beekeeper to pander to this notion than to try to educate his customers as to what constitutes the choicest honey. The chunk honey is also easily and cheaply produced and can be put up for market quickly. Bulk comb, or chunk honey, is usually packed in three, six, twelve or sixty-pound cans. For shipping, the three- and six-pound cans are conveniently crated in cases of sixty pounds each; the twelve and sixty-pound cans in crates of one hundred and twenty pounds.

Regardless of which of the three varieties of honey the beekeeper produces, extracted, section or bulk comb, he must adhere firmly to certain practices if he would obtain the maximum amount of honey. As soon as the white "brace-combs" or bits of wax begin to appear at the tops of the brood-frames, the super should be placed upon the hive. Regardless of whether it contains sections or extracting frames, these should contain full sheets of foundation—or at least starters, which are merely narrow strips of foundation in the frames or sections, instead of full-sized sheets. If possible, a frame of unsealed honey, or a few partially filled sections saved over from the previous season, or taken from another colony, should be placed in each super. These "baits" will induce the bees to enter the supers and commence work much sooner than they otherwise would. If it is desired to keep the queen from laying in some of the super-frames a queen-excluding honey-board (Fig. 16) can be placed between the brood-chamber and super. In case a colony obstinately refuses to go to work in the super, take an entire super—bees and all—from another colony, during the middle of the day and place on the obstinate colony, at the same time transferring the empty super to the other colony.* This exchange of supers will in the majority of cases start them to work in good shape. As soon as the first super is three-fourths filled, lift it up and insert another empty one beneath it. By the time this is half filled, or over, the upper one should be nearly or completely finished and capped. This process of "tiering up" may be continued until the honey flow is nearly over, but care must be exercised towards the last or a number of unfinished combs will remain in the supers. Before being removed from the hive all honey should be sealed. This indicates that it is sufficiently evaporated or "ripened" so that it will keep. If sufficient supers are not on hand to tier up during the entire honey flow, it will of course be necessary to extract from time to time, or remove the completed combs and substitute empty ones in their place. The same will hold true of those localities with an exceptionally heavy flow. In no case should honey be taken off before sealed, as by so doing its keeping qualities will be impaired and it will later turn sour or spoil.

MARKETING HONEY.

For those who are not extensive producers of honey it is best to create and depend upon a local market. Honey is esteemed a delicacy and in every locality some buyers will be found. If the beekeeper will practice strict cleanliness in his work, offer for sale only honey of good flavor, thoroughly ripened, and advertise to some extent he will have no trouble in building up a local trade at fair prices.

When the crop is so heavy that it cannot be disposed of locally, the honey can be sold to firms which make a business of buying and selling honey, or it can be shipped to reliable commission men.

Many extensive beekeepers find it profitable to work up a trade with grocers and others in distant as well as in local towns, thereby combining marketing with production and getting the profits of both.

FRAUDULENT PACKING.

A deceptive method of packing bulk comb honey has recently come to our attention and it cannot be too strongly condemned, both by customers and honest beekeepers. As explained on a preceding page, bulk comb honey, when properly packed, consists of cans filled *full* of *comb* honey, the latter cut into just as large pieces as will go into the can. What few openings *then* remain are filled with extracted honey.

Some beekeepers have, however, adopted the plan of filling the honey cans *only about one-third full* of comb and then filling up the can with extracted honey. Of course the pieces of comb float on top of the extracted honey and when the customer takes off the cover the can *appears* to be filled with comb honey. The deception is not discovered until the customer has purchased the can and used out about a fourth of its contents. Such a deception is little short of actual fraud, for bulk comb honey usually sells at from two to five cents per pound higher than extracted and when the customer pays the

---

*This can be done only when a queen-excluder is used, or after the queen has been definitely located in the lower portion of the hive; otherwise the queen might be carried along in the transfer of supers.

higher price for bulk comb he is certainly entitled to it, not to a mixture containing 75 per cent of a lower priced honey.

BEESWAX.

Beeswax is a product of the bees, and makes up the larger part of the constituents of the comb. When the necessity for comb building occurs in the hive, as at the approach of the swarming season and the honey flow, a greater or lesser number of bees gorge themselves with honey and cluster from the top of the hive or upon a comb, remaining quiet for some time. Presently (varying from one-half to three days, according to different authorities) little scales of wax appear upon the under side of the abdomen, being the resulting secretion of certain glands located at that point. These minute scales are then taken—probably in the majority of cases by other bees—and carried to the newly building cells. Here by being thoroughly worked up and mixed with secretions, they are added to the foundation to make the complete comb, or built directly into comb if foundation is not present. Ordinary comb is therefore made up largely, but not entirely, or pure wax. In working about the apiary all bits of comb, spoiled or broken foundation, burr and brace combs, etc., should be placed in the solar wax extractor. These small savings will in the course of a season amount to considerable, and will more than pay for the time and trouble required. In order to whiten the wax and make it more presentable, it may be placed in the solar wax extractor several times in succession, the sunlight having a marked bleaching effect upon it. Wax is in good demand by all manufacturers of foundation and bee supplies, and commands a ready sale at all times. Wax should never be adulterated in any way. Aside from moral reasons, such adulteration is readily detected and, furthermore, any adulteration renders wax entirely worthless.

SWARMING.

All forms of life, that they may not become extinct, have some method of reproduction. In the case of the greater number of animals this reproduction is merely the reproduction of individuals. In the case of bees it will be seen that the increase of individuals alone would only result in the strengthening or maintenance of the colony. Were reproduction of individuals the only method of increase, the species would ultimately become extinct, for a colony under natural conditions cannot exist indefinitely. In all forms of communistic life in the animal kingdom, we find also a division of communities or colonies. In the case of bees this division takes the form of "swarming," and will be briefly described.

In the spring when the plants are yielding plenty of nectar and pollen, the combs are being rapidly filled with honey and the bees are increasing rapidly within the hive, the bees become possessed of the "swarming fever." Regarding the cause or nature of this but little is understood, except that it is the instinct calling for a division of the commonwealth. The changes that take place within the hive at such times are well understood, however remote the real nature of the swarming fever. At this time the bees evidently prepare specially constructed cells, in which the queen deposits eggs, these eggs differing in no way from those which regularly produce workers. As soon as hatched, the young larvae in these cells are fed by the nurse bees with a special food designated as "royal jelly." Regarding the nature and composition of the royal jelly little is known, except that through its influence and great abundance the sexual organs of the larvae are fully developed during growth and do not remain rudimentary and useless as in the case of the worker bee. The queen-cell is much larger than the ordinary cell, and is elongated so as to form a cone-shaped receptacle, very easily found upon the comb, usually at the lower edge. The queen requires a shorter time for development than does the worker, and during her entire larval and pupal stages is protected by the bees from the old queen who, if permitted, would destroy her. Nine days after the egg is deposited the queen-cell is sealed over and in seven days more the perfect queen emerges. At this time or before, weather conditions being favorable, the old queen and a large portion of the bees in the hive swarm out and seek for another location. Soon after leaving the hive they usually cluster on a limb or bush and remain there for some time, varying from a few minutes to several hours. It is supposed that while thus clustered they send out scouts to find a suitable location, as a hollow tree, wherein the new colony can make its home. This habit of clustering before leaving the vicinity makes it possible for the apiarist to capture the new swarm, although in some cases it departs from the apiary without showing any inclination to cluster. Evidence is at hand, also, that a swarm may wander about for several days or longer, gradually losing in numbers until entirely destroyed.

When clustered upon a limb the colony is easily secured by cutting off the limb and shaking the entire swarm off onto the ground in front of the hive which it is desired it shall occupy. This hive should be prepared beforehand and placed in its permanent location awaiting the swarming of the bees. The frames should all contain full sheets of foundation, wired, and when the swarm is placed in the hive a frame of brood, without bees, should be taken from another colony and placed therein. This will insure the bees remaining in the new hive. The super should also be placed upon the hive at once, as immediately following the swarming experience the bees repair to work with much vigor. A device known as a "swarm-catcher" is sometimes used for taking a swarm from a high limb (Fig. 22).

PREVENTION AND CONTROL OF SWARMING.

The control of swarming is closely coupled with the question of surplus honey production. Inasmuch as the honey-gathering ability of a colony depends primarily upon the number of workers in the hive it will be seen that it is of decided advantage to keep all colonies as strong in numbers as possible. Swarming decreases the number of bees in the colony, and could it be entirely prevented would very materially increase the honey production. Where increase is desired it is advisable to allow each colony to cast one swarm a season, and for the beginner we would not advise the attempt to control swarming by any of the complicated methods often recommended in bee journals. None of them are perfect and considerable loss of time and labor, besides confusion, may result if the inexperienced bee keeper attempts to solve this problem—one which should be attempted only by experienced apiarists.

Upon one important point, however, experienced beekeepers are fully agreed and that is the fact that a crowded condition of the hive is what precipitates swarming. As soon as the hive becomes packed with honey the bees will prepare to swarm, but by furnishing an abundance of super room the beekeeper can defer swarming for a considerable time and sometimes prevent it entirely. If surplus honey, rather than increase of bees, is desired the beekeeper should see that the bees are supplied with plenty of super room during the early part of the season and all during the honey flow.

A few practical methods have been devised by which swarming may be anticipated, as it were, and increase of the colonies made without swarming. One of the most successful of these methods is the one known as the "shaking" method, frequently employed in Texas. Briefly, this is as follows: When the swarming fever is fairly on, the old hive is removed from its stand and a new hive put in its place containing frames with full sheets of foundation. The super, bees and all, is now taken from the old hive, and placed upon the new. The combs are then taken from the old hive and the bees shaken off in front of the new hive. The queen is also placed in the new hive. The old hive, now containing nothing but brood in all stages and a few bees, is placed in a new location, the entrance narrowed down to a small aperture and left to itself. The bees in this hive, hatching rapidly, will care for the brood and will rear a queen from among the young larvae; hence at the end of 21 days it will be a complete colony. Instead of allowing the bees to rear a queen, a mature queen or a "ripe" queen-cell (a cell in which the queen is nearly ready to emerge) can be given this colony. It is supposed that the rough handling, shaking and entering the new hive in a way satisfies the swarming fever of the bees, and upon this point apparently rests the success of the method. If no increase in the number of colonies is desired the shaking process can be modified to meet this requirement. When all the brood in the old hive has hatched, which will be at the end of 21 days after the shaking, the bees can be shaken off the combs in front of the new hive. They will be readily accepted and as the swarming fever will have been overcome in the meantime the colony will now possess its original number of workers. The frames in the old hive may be used in the building up of neuclei or the honey may be extracted from them and the frames used in extracting supers or in brood-chambers of other colonies.

"Artificial division" of the colonies is sometimes resorted to where increase in the number of colonies is especially desired. When the colony is nearly ready to swarm the queen and four or five frames of bees and brood are placed in a new hive and the latter moved to a new stand. The extra space in both hives is filled with frames containing full sheets of foundation. At least one queen-cell is left in the colony occupying the old stand and within a few days this colony has a laying queen. This division may be made even before swarming time so that the young colonies will have opportunity to gain strength before the honey flow. Division should not be made in any case unless the colony is very strong and it must not be made too early in the spring or "chilling" of the brood may occur during cool nights. In short, considerable experience is necessary in order to make this method a success, and the beginner, if he attempts artificial division at all, should try it only when the colony is nearly ready to swarm.

Prevention of swarming by proper manipulation of the "divisible brood-chamber" hive is said to give good results in the hands of experts; but the use of such a hive by the amateur beekeeper is not to be recommended.

Fig. 20—Alley queen and drone trap (after Phillips, Farmers' Bulletin No. 397, U. S. Dept. of Agr.).

Where natural swarming is allowed by the beekeeper, several plans may be resorted to with success to prevent the issuing swarms from leaving the apiary. Among these we may mention clipping the wings of the queen and the use of the drone and queen trap, or the entrance guard. At the approach of swarming time the queen in each hive may be caught and the outer portion of the wing on one side clipped off, care being taken not to cut too close to the body. The queen should always be picked up by the wings or thorax from the combs, whether for clipping or not, and never handled by the abdomen. In clipping the wings hold the queen between the thumb and forefinger of the left hand, grasping her by the thorax. The outer portion of the wing can now be clipped off with a small pair of scissors, and the queen gently replaced upon the comb. In the absence of scissors the wing can be taken off with the blade of a sharp pocket knife, by so holding the queen that the wing to be clipped lies flat upon some smooth wooden surface, as upon the top of the hive, and the edge of the knife blade pressed down upon it. When a swarm emerges from a colony having a clipped queen the queen will attempt to follow but being unable to fly will fall in front of the entrance or near it where she can be readily picked up by the apiarist. The swarm may continue flying in the air for some time, usually long enough for the apiarist to remove the old hive and substitute a new one. The swarm, finding its queen missing, will return to the old stand and enter the new hive placed to receive it. While the swarm is entering, the old queen is released and allowed to enter with it. The hive containing the brood, swarm and young queen, or queen-cell, is placed on a new stand. The device shown in Fig. 20, known as the drone and queen trap, is used in much the same way. As will be noticed from the figure, this trap exactly fits over the entrance to a hive and no bees can enter or leave without passing through the oblong openings in the perforated zinc of which the trap is made. These openings are of such a size that workers may pass through readily, but neither drones nor queens can pass at all.

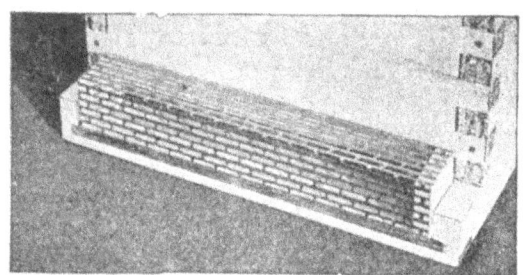

Fig. 21—Entrance guard.

Above the lower part of the trap is a separate compartment, having openings leading into it from below, large enough to admit queen or drones, and covered with cones of wire netting. The queen, in attempting to escape from the hive with the emerging swarm and finding herself unable to pass through the smaller openings, passes into the upper compartment and is trapped. The trap and all can now be removed by the apiarist. The old hive is removed and a new one put in its place as before. When the swarming bees return the queen is released from the trap and allowed to enter with them. This same device is also used for catching the drones from a colony where they are no longer needed for fertilizing queens, or where the drones come of stock which we do not wish to mate with select queens. The entrance guard, shown in figure 21, is used in a similar way, except that it will not trap either drones or queens, although it will prevent them leaving the hive. This is especially useful to prevent the flying of drones from undesirable colonies when the mating of young queens is in process.

UNITING WEAK COLONIES.

It sometimes happens that autumn will find the beekeeper with several weak colonies upon his hands. These may be swarms which issued late in the season or they may result from queenlessness or other causes. It is well to "unite" these into strong colonies so they will withstand the winter in better shape. Nothing is lost by uniting them. Though the number of *colonies* retained by the beekeeper is diminished, he nevertheless still has on hand the same number of *bees* and, after uniting the weak ones, has colonies which will be likely to survive the winter and be in prime condition for work at the beginning of the spring honey flow.

A number of different methods of uniting are explained in textbooks on bee culture, but one which was brought to our attention by Mr. E. G. Le Stourgeon of San Antonio is far and away so much better than any other we know of that we hardly deem it necessary to describe any other. Mr. Le Stourgeon's method of uniting is as follows:

The queen should first be removed from the weaker colony, or if the queen in one of the hives is a poor one, she should be the one removed. The cover is now taken off both colonies and *two* sheets of ordinary newspaper laid *completely* over the top of the strongest colony, so that no openings remain. The brood-chamber of the weaker colony is then lifted bodily from its bottom-board and placed on top of the stronger colony. The hive-cover is placed on top of the uppermost hive-body and is lifted a very small space for a few toothpicks stuck in between the cover and upper edge of the hive-body. We thus have the two colonies on one stand, one above the other, but separated from each other by the two thicknesses of newspaper. The bees on both sides of the newspaper immediately begin cutting and by the time they have cut through the paper (twenty-four to forty-eight hours) they will have become "acquainted" and will not fight. The toothpicks are placed under the cover to give the bees in the upper chamber ventilation until the papers have been perforated. About two days after uniting in this way the beekeeper should open the hive and remove the pieces of paper remaining. We have tried this method several times and in every case the results have been entirely satisfactory.

QUEEN-REARING.

Queen-rearing is a highly specialized branch of the industry and in a bulletin like the present one it is not deemed advisable to discuss the methods in use. The average beekeeper should know, however, what to do in case any of his colonies become queenless through careless handling, accident or otherwise. When a colony becomes queenless and there are eggs or very young worker larvae in the hive the bees will rear a queen themselves without further attention. If, however, they are "hopelessly queenless," that is, have neither brood, eggs nor queen, they can be given a frame of young brood from another colony and they will then proceed to rear a queen. In preference to the above, the beekeeper may purchase a queen from some of the queen-breeders who make a specialty of rearing them. Under present con-

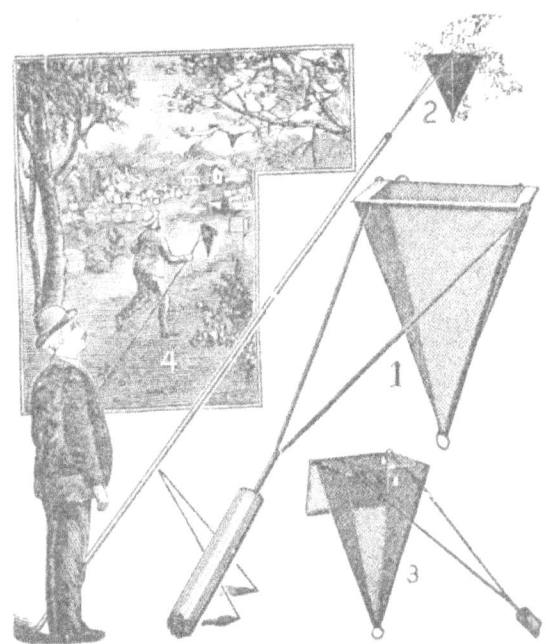

Fig. __.—Adanum swarm-catching device (courtesy of A. I. Root Co.).

ditions, and considering the cheapness of queens, it will not usually pay the amateur beekeeper to rear queens of his own. They can be purchased at about the following prices: Untested, 65c to $1.00; tested, 85c to $1.50, according to the season of the year. An "untested" queen is one which has been fertilized and has already deposited eggs, but which has not been kept sufficiently long to see that her bees, when matured, prove that she has been mated to a drone of the same race as herself. A "tested" queen is on that is not only known to be mated, but is positively known to be purely mated, that is, to a drone of her own race. To distinguish these from a queen that has not yet mated, the latter is designated as a "virgin" queen. The beekeeper who wishes to study queen-rearing in detail, with a view to rearing his own queens, should consult some of the text-books upon bee culture, or books which are devoted exclusively to the subject of queen-rearing.

By consulting the advertisements in the various bee journals, the beginner will find the addresses of queen-breeders throughout the country, together with the races and strains of which they make a specialty. As a usual thing, queen-breeders are found to be reliable. The fact is that the continuance of their business depends largely upon the purity of the queens they ship. Fortunately, none of the bee journals will knowingly accept an advertisement from an unreliable party, and this makes it well-nigh impossible for a fraudulent queen-breeder to remain long before the people.

"LAYING WORKERS."

This is a condition which is most annoying to the apiarist. When a colony has been queenless for a considerable time some of the workers seem to get the idea that they are queens. At any rate they proceed to deposit eggs, but these eggs, being unfertilized, hatch only into drones. The bees in the colony soon get lazy and do little if any work. This condition can be detected by (1) absence of any queen in the hive, (2) the irregular manner in which the eggs are deposited by the laying workers, (3) the bullet-shaped cappings over the drone pupae in the worker-cells and, if the colony has been queenless for a long time, (4) the abnormally large number of drones present in proportion to workers. The bees are also constantly building queen-cells over the male (drone) larvae and trying to make queens of them—a very hopeless ambition indeed.

Colonies having laying workers will very rarely accept a good queen when she is introduced and will sometimes kill as many as a dozen queens in succession. It is usually a waste of time and money to try queening a laying-worker colony. However, one always wishes to save the bees and combs and make use of them. A plan often adopted by Texas beekeepers and the most successful plan we know of, is as follows:

The laying-worker colony is picked up, hive and all, and carried a few hundred yards from the apiary. Here all the combs are taken out and *all* bees shaken or brushed onto the ground. They are left there and the empty hive and its combs carried back to the original location. Most of the workers will get back there before the hive does and practically all of them will be back in the hive in an hour or so. The laying workers (which can not be distinguished from other workers) however, possessing certain queen characteristics, do not find their way back to the hive and are lost. At any rate the method usually works and after this treatment, a queen can be introduced or, better yet, the colony can be united with a good strong colony having a laying queen, by the process just described under the head of "Uniting."

ENEMIES OF BEES.

The beekeeper, like those engaged in all other pursuits connected with agriculture, has enemies to contend with. The enemies of bees and their products fall naturally into two classes; the insect enemies, and the diseases caused by bacteria, moulds or pathological conditions.

THE WAX-WORM.

The wax-worm is the cause of constant and often heavy losses where bees are kept in box hives, or where the colonies are allowed to become weakened. To the progressive beekeeper, however, this pest offers few terrors.

The adult insect is a small brownish moth that may be seen flitting about empty combs, abandoned hives and similar places, always searching for a place to deposit her eggs. The latter are laid by preference upon the combs themselves but are often laid in cracks and crevices of the bee hives or in particles of beeswax or stored honey kept in the extracting house or other building. The eggs hatch into small, ash-colored worms which, when mature, are about an inch in length. These worms eat their way through the combs, destroying any bee larvae that may be in their way. They spin a silken web as they go, and a badly infested comb soon becomes a filthy mass of worms, excrement and webs.

The first requisite in fighting these pests is to have the bees in nothing but good, tight, modern frame hives. Nothing can be accomplished against this nuisance where bees are kept in box hives or "gums." In a frame hive all combs can be taken out by the beekeeper and examined for the worms. When they are detected they can be cut out of the combs with the point if a knife and destroyed. The silken galleries through the combs at once indicate their whereabouts. In addition to destroying the worms found in the combs, the bottom-boards of the hives should be kept cleaned of all trash and cuttings.

The second requisite is to have all the colonies strong, that the bees themselves may fight the moths successfully. Italian bees are rarely troubled by the wax-worm, and in localities where the wax-moth is prevalent the beekeeper should Italianize his bees, that is, he should secure Italian queens from some reliable queen-breeder and introduce them into his colonies. This is not an expensive or difficult process. It is only necessary to purchase the queens and, upon their arrival, follow the directions for introducing which are furnished by all queen-breeders.

Every particle of wax and comb that is left lying about, either out of doors or within, furnishes a breeding place for the moths—a source of supply from which moths may go to the hives containing bees and there continue their increase and damage. It is therefore of prime importance that the utmost cleanliness be maintained by the beekeeper in all his work and no breeding places left in which the moths can increase. The wax-worm is especially liable to attack stored comb honey, empty combs or "bait" combs being kept over from one season to another. Even when such materials are kept in tight supers or boxes—too tight for the moths to get at them—they will become infested by the "worms" or larvae. This is probably brought about by the wax-moth depositing her eggs on the outside of the receptacle and the minute larvae crawling inside soon after hatching from the eggs. Stored combs should be examined from time to time and at the first evidence of attack should be fumigated with bisulphide of carbon ("high-life"). We have found it necessary during the summer months to examine combs of this kind at least once a week and during the winter season at least once in every two weeks in order to prevent serious damage by the larvae. Fumigation of combs or comb-honey with the bisulphide of carbon is by no means difficult or expensive and the flavor of the honey is not impaired by the use of this chemical. Supers or hive-bodies containing the combs to be treated are tiered one above another, the bottom one resting on a perfectly tight floor or level board free from cracks. On top of the tier an empty super is placed and all cracks between the hive-bodies made as nearly air-tight as possible. A saucer containing a few tablespoonfuls of the bisulphide is now placed on *top* of the highest row of frames and the tier covered over with a heavy blanket and allowed to remain for several hours. It should be borne in mind that the carbon bisulphide is highly inflamable, and under some conditions explosive, hence care must be taken in its use and handling to avoid proximity to fire of any kind, lighted lamps, lanterns, pipes, stoves, etc. A better plan during summer, when combs are found to be infested, is to place them over a strong colony of Italian bees. The latter will make quick work of the moth larvae. In short, this is the best place to keep such combs, supers, etc., at all times. Strong colonies, in good frame hives, are rarely if ever attacked by the wax-moth.

To summarize the means of dealing with the wax-moth:

Keep bees only in modern frame hives.

Keep all colonies strong and vigorous.

Italianize all black and hybrid colonies.

Keep bottom-boards free of trash and dirt.

Allow no bits of unprotected wax or comb to accumulate on the premises.

ANTS AND OTHER ENEMIES.

Ants of various kinds, and especially the smaller species, often visit the bee hives for the purpose of stealing honey, and some of the more pugnacious ants will even attempt to kill and carry away the bee larvae. As a general thing strong colonies will resist the attacks of the ants successfully although there is one species, known as the "Argentine ant," in Louisiana which attacks the bee hives in such overwhelming numbers that resistance is entirely useless. This particular ant is not known to occur in Texas, but may make its appearance here at some future time.

Very often the ants can be followed to their nests and there destroyed by the use of bisulphide of carbon. It is only necessary to make a hole in the ants' nest with a sharp stick, pour in a little of the bisulphide and cover the entire nest with tightly packed earth. A little water sprinkled over the nest immediately after the dirt is tamped down will add to the effectiveness of the treatment. The explosive nature of the bisulphide must never be overlooked, as explained in our preceding paragraph on the wax-worm. Gasoline, or even boiling hot water, may also be used for destruction of the ant colonies.

The use of "ant poisons" about an apiary is not advisable for the reason that the bees are as likely to be killed by them as are the ants.

In addition to destroying the ants' nests, one can keep the little pests out of the bee hives by placing the latter on stands or elevated benches, the legs of which rest in cups containing water, kerosene oil or crude petroleum ("Beaumont oil"). It is often sufficient to merely rub the legs of the benches with the crude oil, the odor alone being sufficient to prevent the ants from passing. Another excellent expedient is to dip a few bricks in crude oil and place one of these under each corner of the hive, taking care to see that no grass or weeds touch the hive to afford a passageway for the ants.

The Rev. Albert Biever, an enthusiastic beekeeper of New Orleans, finds that vaseline, mixed with a small quantity of kerosene, is very effective when placed about the legs of the hive-stands and repels the ants successfully.

The writer discovered, while keeping bees in Louisiana, that the medium and small-sized ants will not cross a surface having upon it a very small amount of *dry* corrosive sublimate. Strips of cotton tape or of thin cloth were accordingly soaked in a saturated water solution of corrosive sublimate and afterwards dried. These, when wrapped around table legs, etc., effectually prevented the ants from crossing, and we used them in the same way with success in keeping ants away from the bee hives. The tape, however, is not effective when wet nor will it retain the corrosive sublimate after being washed by rains. Its use is not adapted to localities where rains are frequent but the method might well be used with success during the dry seasons in the southwestern part of Texas. In preparing and handling the tape it must always be remembered that the corrosive sublimate is a most powerful poison and every precaution against accidental poisoning must be taken. Particularly must the hands be well washed after making or handling the tape. The use of this poisonous tape under bee hives seems not to affect the bees in any way and it does not deter them from crawling up the legs of the hive-stand when, perchance, they fall to the ground themselves.

Large robber-flies often capture worker bees and perhaps queens, but there is little the beekeeper can do to prevent their work.

Spiders occasionally spin webs near the entrances of the hives and capture a considerable number of the bees. The best preventive of such damage is to leave no cracks or openings about the hive in which the spiders may find a hiding place. Webs should be destroyed when found and the spiders themselves killed at every opportunity.

BIRDS.

Very rarely an individual bird, especially the bird known as kingbird or bee-martin, will capture the bees as they fly to and from the hive. This habit is not a universal one with any bird and is only developed by individuals. Where the offender persists in his attacks a shotgun is the best remedy but in no case should warfare be made upon birds as a whole, or even upon the kingbirds as a rule, to atone for the offenses of a single individual. The kingbird is preeminently, as are to a great degree all of our song birds, insectivorous, that is, its main diet is made up of worms, larvae and insects. The good that birds of any species—barring, possibly, the turkey buzzard—do to the farmer, planter and fruit grower far more than offsets the occasional damage. Their protection and preservation is, therefore, to the utmost interest and welfare of the farmer and fruit grower, as well as to the State at large.

DISEASES OF BEES.

Two very dangerous diseases of honey bees occur in the United States, known respectively as "American foul brood" and "European foul brood." Both are highly contagious and are capable of destroying an entire apiary in a remarkably short time. Other bee troubles, such as paralysis, pickled brood, etc., are of less common occurrence, and, being apparently non-contagious, are far less serious than the two kinds of foul brood, though they occasionally do considerable damage locally.

AMERICAN FOUL BROOD.

This is a disease affecting the larvae, or young bees, before they reach maturity. It is caused by a minute organism, known as *Bacillus larvae*, which multiplies rapidly within the bodies of the bee larvae and causes their death. It cannot originate spontaneously, or of its own accord. It can occur only as a result of the bees obtaining access to hives where the disease occurs, or to honey which has been taken from an infected hive and which itself contains the germs of the disease. Likewise the disease can not disappear of its own accord, but will continue to attack colony after colony until cured by treatment or until the bees are all dead.

The bee larvae are attacked by the disease shortly before time for them to be sealed over. At first the larvae turn to a chocolate color and then to a dark brown. After the larvae are sealed decay continues and the crappings become sunken or perforated. If a toothpick or other sharp piece of wood be inserted into the body of a decayed larva, twisted about and slowly withdrawn, the decayed substance adheres to it and comes out in a string or ropy mass, sometimes as far as one-half to three-fourths of an inch. After the decayed larvae dry they form brown scales in the cells of the comb, these scales being readily detected by holding the comb so that strong sunlight strikes into the cells. In colonies badly diseased a disagreeable odor is always noticed, one that is frequently described as the odor of heated glue. The adult bees in a diseased colony make little attempt to clean out the decaying bodies of the larvae, soon become discouraged and work with lessened vigor. This disease occurs in many parts of Texas and is at present doing enormous damage to the beekeeping industry.

STATE CONTROL AND ERADICATION OF FOUL BROOD.

Chapter 126 of the General Laws of Texas, Twenty-eighth Legislature, provides for eradication of foul brood in the State of Texas and charges the State Entomologist at the Agricultural and Mechanical College with the enforcement of this law.

At the present time a vigorous effort is being made by this office to prevent further spread of American foul brood within the State and to eradicate it where it already occurs. Unfortunately, the disease has obtained such a start in some portions of the State that it is impossible for the State Entomologist, with the appropriation available, to carry on the work in more than a few counties at a time. Material progress has been made in several counties in reducing and eliminating the disease and the work will be undertaken in additional counties as soon as funds permit. The beekeepers in the counties concerned have been generous in their co-operation and assistance and the support of this work in all localities is earnestly requested.

A booklet has been recently issued by this office, in which is contained the "fould brood law" of Texas, regulations for facilitating the eradication of disease and particulars concerning the counties where this work is now being carried on. This booklet is free upon request.

All correspondence relative to foul brood eradication, requests for this booklet, as well as requests for information relative to apicultural matters should be plainly addressed to the "State Entomologist, A. & M. College, College Station, Texas.

TREATMENT FOR AMERICAN FOUL BROOD.

Where only a few colonies, out of many, are diseased, it is by far the safest plan to burn up the diseased colonies, hives, brood, bees and all. To do this stop up the entrance of the hive after dark, so no bees can escape. Bore a one-half inch hole through the cover and pour into the hive about a teacupful of gasoline, afterwards stopping up the opening. *In doing this, do not have a light or fire near enough to explode the gasoline.* This will kill the bees within the hive in a couple of hours. Next dig a hole about six inches or a foot deep, and large enough to more than hold the diseased colonies. In this hole place plenty of kindling and wood. Move the diseased hives to it, open them and pile bees, frames and hives on top of the kindling, adding coal-oil if desired. Be careful that no bits of honey or comb fall outside the hole or out of reach of the fire. Set fire to the entire lot, being careful to avoid the sudden blaze which will be caused by presence of the gasoline. After the burning is completed, cover the ashes with dirt, so as to prevent other bees from getting any drops of infected honey which may not have burned up. The moving and burning of the hives should take place at night, so that bees from other colonies will not get at the infected material.

Where entire apiaries, or a large percentage of the colonies in them, are diseased, treatment is preferable to burning. To treat diseased colonies, proceed as directed below:

As foul brood is a disease caused by a germ, it is necessary to disinfect hands, tools, etc., to kill the germs that get on them in the course of handling infected hives, honey or bees. Secure from your druggist a solution of corrosive sublimate, made at the strength of one part corrosive sublimate to two hundred and fifty parts of water. This is a deadly poison, but there is no danger connected with its use if you use proper precautions. Do not get any of the solution in your mouth, and do not fail to wash your hands well with soap and warm water after using this solution. Keep it labeled, and do not allow children to get at it. The corrosive sublimate solution should be kept in glass, wooden or earthenware vessels—never in metal vessels or bottles. This solution is for disinfecting tools, veils, etc.

Take all hive-bodies, bottoms, covers, super-bodies and tools and store them in a bee-proof house. Take all old brood-frames, used super frames, used combs and particles of wax, and burn them up, following the directions for burning given above. In other words, clear the apiary of all bits of comb, old frames, etc., so as to get rid of all contaminated material. The bottoms, hive-bodies, covers and empty super-bodies should be disinfected in the following manner:

1. Char them well, inside and out, by holding over a fire or by using a gasoline plumber's torch; the latter way is preferable.

2. Then paint them, inside and out, with a solution of carbolic acid, one part of acid to two of water.

Inspect all colonies and mark those which are diseased. In making this inspection, examine, *first*, those colonies which you do *not* suspicion of being diseased. After examining a diseased colony, disinfect hands, tools, veil and smoker by dipping all in the corrosive sublimate solution mentioned above. Of course the smoker must be allowed to cool before being dipped in the solution. As soon as the things are dry, they are again ready for use. You will see at once that your work will be facilitated by having several complete outfits of veil, smoker, etc., so that you can be using one outfit while another is being disinfected.

When making these examinations, or when working with the bees for any purpose, do not allow any drippings of honey, brace-combs, etc., to lie about even for a few minutes. When examining the frames, hold them *over the hive*, so that any honey dripping out will fall back into the hive and not on the ground. Remember that honey from a diseased hive contains the germs of the disease; if bees from other colonies get at it, they will carry the disease to their own hives.

In the case of colonies that have died out entirely, burn the combs and frames and disinfect the hive-bodies, as directed above. Burn all dead bees also.

In the case of weak, diseased colonies, unite two, three or more colonies, burning up the extra frames and disinfecting the hives, bottoms and covers in the same way.

If diseased colonies still contain enough bees so that they are able to continue work, or if they are only fairly strong, stimulate brood-rearing by daily feeding, in small amounts, of syrup made from cane sugar and cold water. Do not feed honey. If your bees were diseased during the last honey flow, your honey contains the germs of the disease, and by feeding it you will only be aggravating the disease. Feeding with sugar syrup will permit some of the young bees reaching maturity, and will result in building up the colony to some extent, despite the disease.

Provide yourself with as many new hives (or disinfected hive-bodies, covers and bottoms) as you have diseased colonies. Secure new brood-frames sufficient for these hives. Also, secure enough foundation for the frames in these hives. Use your own judgment as to using starters, half-sheets or full-sheets.

Nail up the frames, insert the foundation, put them in the new or disinfected hives, and have them all ready for use. You will then be prepared for the treatment proper, which should be made on a day during the height of the honey flow, when all the bees are very busy gathering honey in the fields.

Go to one of the diseased colonies, taking with you one of the new hives containing new frames and foundation. Place the hive containing the diseased colony to one side and put the empty hive in its place. Take the frames from the old hive, one at a time, and shake or brush the bees off them in front of the empty hive. After the bees are shaken off, return the brood-frames to their own (old) hive. You will thus get all adult bees and the queen (but no brood) into the new hive. As soon as the shaking is completed, take the hive of brood and put it where the bees can not get at it.

If new honey is in the combs at the time of this operation, the bees should be *brushed* from the combs and *not shaken,* as shaking dislodges the honey, and it is desired to retain every drop of this in the old comb.

Proceed in the same manner with the next diseased colony, and so on until all the diseased colonies have been treated. The disease should not appear again in these treated colonies unless bees from them obtain access to infected colonies or infected honey.

If one wishes to make doubly sure of success, the treatment can be repeated three days later, the bees being again shaken onto new frames containing fresh foundation, and the combs, built in the meantime, destroyed.

After completing the treatment, you will have on hand hives full of diseased brood. This diseased brood is exceedingly dangerous, for it is literally teeming with billions of the foul brood germs. Some beekeepers tier up the hive-bodies containing such brood, over a weak colony, allow the bees to mature (that is, as many as are not killed by the disease), and then treat this colony, afterwards burning up the empty brood-frames or rendering the combs into wax. However, we personally prefer to burn up these frames of diseased brood as soon as the shaking process is over. We consider it much safer. After burning up the brood-frames, the hives which contained them should be charred and disinfected as directed above.

It is a very difficult matter to destroy the germs in infected honey. Ordinary boiling will not kill the spores of the disease. The only way in which the honey may be made sterile (free from germs) is to boil it for at least an hour in a *tightly closed vessel.* Even then there is danger of a few spores remaining in the honey, and we do not consider that it is economy to take such a risk of perpetuating the disease in order to save a small amount of honey.

When combs from diseased colonies are rendered into wax, it should be done by boiling in water or, better still, in a steam wax-extractor. Never render such combs in a solar extractor.

If you have a large apiary, and only a very few colonies are diseased, we would advise burning these up *in toto* at once, instead of treating them. This would not be economical, however, if a considerable number of colonies were infected.

After completing the treatment, and disinfecting everything, keep a close lookout for reappearance of the disease, especially after the honey flow subsides and the bees get to robbing as much as they can. Treat diseased colonies as fast as they are located, but when there is no honey flow to speak of treat only at dusk in the evening, so there will be no robbing. When there is a honey flow on, and the bees are very busy, the shaking treatment may be carried on all day without interruption or danger.

Treatment should always be discontinued immediately should the bees commence robbing.

EUROPEAN FOUL BROOD.

This disease was formerly known to beekeepers under the term of 'black brood." More or less confusion has existed as to the identity of "black brood," but, through the researches of Drs. Phillips and White, of the Bureau of Entomology, United States Department of Agriculture, it has been quite well established that many, at least, of the cases of "black brood" were of the disease now designated as "European foul brood." While not quite as destructive as "American foul brood," is it, nevertheless, almost as much to be dreaded.

European foul brood has not yet been found in Texas, but is known to occur as near as certain counties in Arkansas, and every Texas beekeeper should be on the alert for its appearance here, especially as there is always the possibility of its being introduced with shipmnts of bees from other States. The following description is taken from Circular No. 79, Bureau of Entomology, by Dr. E. F. Phillips, and will be of use to beekeepers in distinguishing this disease from American foul brood:

"Adult bees in infected colonies are not very active, but do succeed in cleaning out some of the dried scales. This disease attacks larvae earlier than does American foul brood, and a comparatively small percentage of the diseased brood is ever capped. The diseased larvae which are capped over have sunken and perforated cappings. The larvae when first attacked show a small yellow spot on the body near the head and move uneasily in the cell. When death occurs they turn yellow, then brown, and finally almost black. Decaying larvae which have died of this disease do not usually stretch out in a long thread when a small stick is inserted and slowly removed. * * * There is very little odor from decaying larvae which have died of this disease, and when an odor is noticeable it is not the "glue-pot" odor of the American foul brood, but more nearly resembles that of soured dead brood. This disease attacks drone and queen larvae very soon after the colony is infected. It is, as a rule, much more infectious than American fould brood, and spreads more rapidly. On the other hand, it sometimes happens that the disease will disappear of its own acord, a thing which the author never knew to occur in a genuine case of American foul brood."

The treatment for European foul brood is the same as that already outlined for American foul brood.

PICKLED BROOD.

A peculiar affection of brood, known to beekeepers generally under the name of "pickled brood," sometimes appears suddenly and does a very considerable amount of damage before being checked. During the past spring (1911) it was more prevalent and destructive in Texas apiaries than ever before known.

In this disease the larvae die at various stages, but usually when half grown or larger. In severe cases, many larvae die after being sealed over and the dead sealed brood has much the apearance of having been scalded with a dash of hot water. The cappings are rarely perforated to any considerable extent. The dead larvae turn dark or dark yellow soon after death. The dead bodies do not disintegrate quickly, as in American foul brood, nor is the "ropiness" present. The characteristic odor of American foul brood is lacking and, instead, the brood has a very acrid, sour smell. In the first stages of the disease, and during mild attacks of it, the bees persistently remove the dead larvae, but seem unable to keep up with this task when the trouble becomes real severe. This trouble is hard to distinguish from European foul brood and, on the other hand, it is sometimes hard to tell from chilled or overheated brood. One remarkable difference is noted, however. If a frame of pickled brood is placed in a healthy colony the disease is almost immediately communicated to the healthy brood on either side of it. The same thing occurs when a frame of healthy brood is inserted in a colony having pickled brood; the healthy brood becoming infected very quickly. Such a "contagion" does not occur when chilled brood is placed in the same hive with healthy brood. Some authors go so far as to express the opinion that pickled brood is nothing more than brood which has died as a result of chilling, under-feeding or other natural cause, but in view of our own observations and experience with it we can hardly subscribe to such a view.

By a number of authorities the view is held that pickled brood is the manifestation of a pathological condition, or a "constitutional" disease, it being held that the queen is in some way incapable of producing eggs which can develop into healthy larvae. Experience lends considerable support to this theory, as a number of leading beekeepers have found that a cure of the trouble has followed a change of queens. In our experimental apiary the past spring we had but one colony affected with pickled brood, but the case was a very severe one. A number of treatments were tried without success, and finally the queen of this colony was killed and a strong, vigorous queen from a healthy colony introduced in her stead. Although nearly all brood in the hive was dead at the time of her introduction, the dead brood was rapidly cleaned out and the larvae hatching from the new queen's eggs were all healthy. In the course of a month after her introduction all signs of the disease had disappeared.

The outcome of this one experiment would not, of course, justify us in stating that re-queening will always cure the disease, but our experience is corroborated by the similar experience of other beekeepers, and in cases of pickled brood it may be well worth the while of the beekeeper to give this method a trial.

CHILLED, STARVED, OVERHEATED BROOD, ETC.

Not infrequently brood gets too cool and is killed as a result. Normally, the bees will cluster thickly over the brood and keep it warmed to the proper temperature, but cold nights, coming late in the spring, may lower the temperature so that the bees are not able to "hover" all brood in the hive at a time when considerable brood is present. Under such circumstances the brood in the outer frames is killed. The same thing may happen after the beekeeper has divided his colonies for the purpose of making increase, and has failed to reduce the entrances sufficiently to enable the bees to maintain the proper temperature. Brood killed by chilling is usually removed rapidly by the bees, particularly in colonies that are fairly strong. Chilling can be almost entirely prevented if the beekeeper will avoid too large entrances, especially as long as the nights are cool. Excessive "spreading" of the brood is also to be avoided.

Death of the larvae sometimes results from just the reverse condition—too high a temperature and insufficient ventilation. If entrances are of fair or liberal size, overheating of the brood is not likely to occur except in the very hottest weather, and not then unless the hives are unshaded. Suffocation of bees and brood is most likely to occur when bees are shipped with insufficient openings for ventilation.

A sudden stoppage of the honey flow, when brood is abundant and the colonies low in stores, may often result in larvae dying for lack of sufficient and proper food. The remedy for such a condition consists of immediate feeding to meet the needs of the colonies.

The statement is made by some authors that brood is at times poisoned as a consequence of fruit growers spraying fruit trees while in bloom, the nectar being poisoned and the workers bringing it to the hives. It seems more probable that the poisons ordinarily used for spraying would be likely to kill the majority, if not all, of the workers attempting to carry the poisoned nectar. Nevertheless, when dead brood occurs and can not be traced to any other definite cause, it may be well to ascertain whether any such spraying has been done in the neighborhood.

In Louisiana we have noticed a brood trouble which does not seem to have been heretofore recognized. With continued damp weather, mildews and moulds find lodgment in the combs of colonies not located in open, well ventilated places. The fungi mentioned find a favorable medium in the pollen stored in the cells and perhaps also in the unsealed honey and in the semi-liquid food placed by the nurse bees about the larvae. In such colonies considerable dead brood is found and the masses of fungi growing on the cells of pollen are much in evidence. The dead brood is brownish in color and watery, while the odor from affected hives is very strong and in no way distinguishable from the "glue-pot" odor accompanying American foul brood. The decaying larvae do not show the "ropiness" of American foul brood and the fact that a return of dry or cool weather always causes the trouble to disappear entirely refutes any opinion that it is a contagious or bacterial disease. While we have not seen this affection in Texas it is nevertheless quite likely to appear in localities of heavy rainfall or during protracted periods of wet weather.

PARALYSIS.

The diseases thus far considered affect only the brood or larvae; never the adult bees.

But one disease is known which affects the adult bees and this is termed "paralysis." It does not affect the larvae. The workers affected have their abdomens somewhat swollen and the entire body assumes a dark, greasy appearance. Badly affected workers shake or tremble quite violently. The affected workers are often noticed at the entrance or on the ground about the entrance after their eviction by the healthy members of the colony.

The editors of "A. B. C. of Bee Culture," Messrs. A. I. and E. R. Root, recommended as a remedy the exchanging of diseased and healthy colonies. That is, the colony affected with paralysis is placed on the stand occupied by a strong, healthy colony and the latter is placed on the stand formerly occupied by the colony with paralysis. The resulting addition of strong, vigorous bees to both colonies is thought to result in the eviction of all diseased bees and the infection is thus eliminated and a cure effected.

EDUCATIONAL.

The beekeeper who, by experience with his bees, has verified the simple statements made in this bulletin will doubtless wish to pursue his studies further. For this purpose he may secure any one or more of a number of excellent text books on apiculture, among which we may mention the following:

Comstock, Anna B.—"How to Keep Bees."
Root, A. I. & E. R.—"A. B. C. and X. Y. Z. of Bee Culture." (In French, English or German.)
Maeterlinck, Maurice—"The Life of the Bee."
Miller, C. C.—"Fifty Years Among the Bees."
Cook, A. J.—"Manual of the Apiary."
Hutchinson, W. Z.—"Advanced Bee Culture."
Buschauer, Hans—"Amerikanische Bienenzucht," (in German).
Doolittle—"Scientific Queen-Rearing."

Among the magazines devoted exclusively to beekeeping "Gleanings in Bee Culture," published at Medina, Ohio, and the "American Bee Journal," published at Chicago, Ill., are perhaps best suited to Texas beekeepers as both have departments regularly devoted to beekeeping in the South.

Beekeepers who keep even a few colonies should not fail to join both the State and National Beekeepers' Associations, both of which are devoted to furthering the interests of the beekeeping industry. Full particulars regarding both these Associations may be had free of charge by addressing their respective secretaries. Mr. E. B. Tyrrell, 230 Woodland Ave., Detroit, Mich., is Secretary of the National Association, and Mr. W. T. Childress, Batesville, Texas, is Secretary of the Texas Beekeepers' Association.

The amateur will also gain considerable information by requesting catalogs of the various firms which advertise in the bee journals.

Where special problems arise, when difficulties are encountered or when specific information is wanted upon any phase of beekeeping the State Entomologist's Department at College Station, Texas, may be consulted. All such letters will be cheerfully answered at all times and without charge.

College Station, Texas.
December 18, 1911.